ANALYSIS OF
WATER DISTRIBUTION
SYSTEMS

ANALYSIS OF WATER DISTRIBUTION SYSTEMS

Thomas M. Walski

U.S. Army Corps of Engineers, Waterways Experiment Station
Vicksburg, Mississippi

VNR VAN NOSTRAND REINHOLD COMPANY
—————————————————————— New York

Copyright © 1984 by Van Nostrand Reinhold Company Inc.

Library of Congress Catalog Card Number:
ISBN: 0-442-29192-2

Manufactured in the United States of America

Published by Van Nostrand Reinhold Company Inc.
115 Fifth Avenue
New York, New York 10003

Van Nostrand Reinhold Company Limited
Molly Millars Lane
Wokingham, Berkshire RG11 2PY, England

Van Nostrand Reinhold
480 La Trobe Street
Melbourne, Victoria 3000, Australia

Macmillan of Canada
Division of Canada Publishing Corporation
164 Commander Boulevard
Agincourt, Ontario M1S 3C7, Canada

15   14   13   12   11   10   9   8   7   6   5   4   3   2

Library of Congress Cataloging in Publication Data

Walski, Thomas M.
    Analysis of water distribution systems.

    Includes bibliographies and index.
    1. Water—Distribution.   2. Water-supply engineering.
I. Title.
TD481.W34   1984   628.1'44   84-3554
ISBN 0-442-29192-2

To my parents, to whom I owe so much.

PREFACE

This is the book I wish someone else had written before I finished school. Such a book would have saved me a lot of time and trouble. Instead I have had to learn things the hard way—by making mistakes and doing things over until I felt comfortable in what I was doing. So I am writing this book to share what I have learned and to spare others a lot of the difficulties I have encountered.

Water distribution systems have been around for many years and engineers have been studying them the entire time. Unfortunately, the engineers with utilities and consulting firms do not have the time to document what they have learned. So, much of what is known about the subject is carried around in the heads of practicing engineers, never to be published. Young engineers must learn the tricks of the trade from other engineers or by piecing together information from their own experience and the literature. This book documents techniques for solving many of the problems faced by engineers. It is hoped that it will make life somewhat easier for the young engineer.

This is not the first book to be written on this subject. American Water Works Association (AWWA) manuals and Texas Water Utility Association manuals provide a good deal of guidance, but much of the information is not quantitative. For example, AWWA Manual M8, *Water Distribution Training Course,* describes the need for cleaning and lining water mains and how the work should be done, but does not give quantitative rules for selecting the mains to be cleaned and lined. This book provides such guidance.

Other works, such as Buettner's notes from his *Practical Hydraulics and Flow Monitoring Workshop* and Jeppson's *Analysis of Flow in Pipe Networks* provide excellent information on certain individual topics, namely, testing and the mathematics of solving network problems. This book is intended to be more general than these very helpful works.

One topic that is conspicuous by its absence is waterhammer (hydraulic transients). Unfortunately, I do not feel that I have enough practical experience on the subject to contribute anything original, so I will save this topic for the second edition of this book. By then I will hopefully be able to write something meaningful on the subject.

I am serious when I mention a second edition of this book, as I do not feel the present effort is the definitive work on water distribution systems. Rather, I see it as a summary of the state of the art which will hopefully be out of date in the next decade because of all the advances made in this field.

In keeping with the thought that this book is not the source of all wisdom in the field, I welcome any feedback from readers. If someone should have a better way to solve a problem that I have presented in this book, please let me know so that I can learn from it. Write me in care of the publisher or at the Waterways Experiment Station in Vicksburg, MS.

I would also like to thank all of those who have helped me learn the lessons I am sharing with the readers of this book. I especially want to thank Dr. Joe Miller Morgan of Auburn University, who reviewed the first draft of the report, and whose comments have made this book considerably more readable. Mr. Tommy Schaefer did a considerable amount the original drafting work for this book, and Mrs. Cheryl Lloyd assisted in proofreading.

I hope this book will help you do your job better.

Tom Walski, Ph.D., P.E.

CONTENTS

Preface / vii

1. Introduction

    1.1. The Problem / 1
    1.2. Overview / 2
    1.3. Distribution System Components / 2
    1.4. Units / 3
    Review Questions / 4

2. Review of Closed Conduit Hydraulics / 5

    2.1. Overview / 5
    2.2. Continuity Equation (Integral Form) / 6
    2.3. Continuity Equation (Differential Form) / 9
    2.4. Momentum Equation (Integral Form) / 11
    2.5. Momentum Equation (Differential Form) / 13
    2.6. Energy Equation / 17
    2.7. Velocity Distributions / 20
    2.8. Calculating Head Loss / 24
    2.9. Head Loss in Laminar Flow / 26
    2.10. Head Loss in Turbulent Flow / 27
    2.11. Friction Factor for Smooth Flow / 29
    2.12. Friction Factor for Rough Flow / 30
    2.13. Friction Factor for Transition Flow / 30
    2.14. Explicit Formulas for Determining Friction Factor / 32
    2.15. Empirical Head Loss Equation / 33
    2.16. Hazen–Williams Equation / 35
    2.17. Minor Losses / 42
    Review Questions / 46
    Problems / 46
    References / 47

3. Solving Pipe Network Flow Problems / 49

   3.1. Introduction / 49
   3.2. Types of Problem / 50
   3.3. Types of Network / 50
   3.4. Energy Equation for Looped System / 51
   3.5. Equivalent Pipes / 52
   3.6. Pumps in Water Distribution Systems / 56
   3.7. Combining Pumps / 62
   3.8. Pump Efficiency / 64
   3.9. Time Dependence / 66
   3.10. Networks with Loops or Multiple Constant Head Points / 70
   3.11. The Flow (Q) Equations / 73
   3.12. The Node (H) Equations / 74
   3.13. The Loop (ΔQ) Equations / 75
   3.14. Numerical Solution Techniques / 76
   3.15. The Linear Theory Method / 77
   3.16. The Newton–Raphson Method / 80
   3.17. The Hardy–Cross Method / 83
   Review Questions / 85
   Problems / 86
   References / 90

4. Using Water Distribution System Models / 91

   4.1. Introduction / 91
   4.2. Selecting the Approach and Model / 93
   4.3. System Maps / 96
   4.4. Initial Estimates of Pipe Roughness and Water Use / 98
   4.5. Calibration Overview / 99
   4.6. Data Collection for Calibration / 102
   4.7. Development of Adjustment Factors for Calibration / 103
   4.8. Implications of Calibration Techniques / 108
   4.9. Production Runs of the Model / 113
   Review Questions / 115
   Problems / 116
   References / 116

5. Sizing Water Mains / 117

   5.1. Introduction / 117
   5.2. Classification of Problems / 118

5.3. Costs / 119
5.4. Pumped Systems / 122
5.5. Gravity Systems / 127
5.6. Rules of Thumb and Nomograms for Pipe Sizing / 136
5.7. Mathematical Programming Approach to Pipe Sizing / 142
5.8. Design Criteria / 143
Review Questions / 145
Problems / 146
References / 147

6. Providing and Restoring Carrying Capacity / 149

6.1. Introduction / 149
6.2. Loss of Carrying Capacity / 150
6.3. Cleaning and Lining Process / 153
6.4. Economic Evaluation of Cleaning and Lining / 156
6.5. Costs / 159
6.6. Comparison with Energy Cost / 160
6.7. Comparison with Parallel Pipe / 166
6.8. Comparison of Lining vs. Cleaning Only / 175
Review Questions / 176
Problems / 177
References / 178

7. Pipe Breaks and Water Loss / 179

7.1. Introduction / 179
7.2. Causes of Breaks and Leaks / 180
7.3. Leak Detection / 182
7.4. Evaluation of Leak Detection / 187
7.5. Repairing Breaks and Leaks / 191
7.6. Projection of Break Rates / 194
7.7. Costs of Breaks and Remedial Measures / 196
7.8. Economic Evaluation of Breaks / 200
7.9. Break Information Record Keeping / 204
7.10. Lost and Unaccounted-for Water / 206
7.11. Appurtenances / 208
Review Questions / 208
Problems / 209
References / 210

8. Testing Water Distribution Systems / 211

8.1. Introduction / 211
8.2. Pressure Measuring Devices / 212
8.3. Differential Pressure / 214
8.4. Pressure Snubbers / 219
8.5. Flow Measurement / 220
8.6. Venturi Meters / 220
8.7. Orifices and Nozzles / 221
8.8. Flow Tubes / 226
8.9. Elbow Meters / 226
8.10. Propeller, Turbine, and Paddlewheel Meters / 228
8.11. Vortex Shedding Meters / 229
8.12. Ultrasonic Meters / 229
8.13. Electromagnetic Flowmeters / 231
8.14. Pitot Tubes / 231
8.15. Measuring Discharge / 240
8.16. Distance / 242
8.17. Pipe Diameter / 243
8.18. Elevation / 246
8.19. Calculations Using Test Results / 247
8.20. Pipe Roughness and C Factor / 247
8.21. Indirect Methods of Determining Roughness / 250
8.22. Hydraulic Gradient Tests / 255
8.23. Fire Flow Tests / 257
8.24. Assessing Capability for Expansion / 262
8.25. Testing Large Meters / 263
Review Questions / 266
Problems / 268
References / 270

Appendix. Conversion Factors / 271

Flow / 271
Kinematic Viscosity / 271
Pressure / 272
Power / 272
Length / 272

Index / 273

1 | INTRODUCTION

1.1. THE PROBLEM

In attempting to understand the behavior of a water distribution system, an engineer is in a position similar to that of a doctor trying to understand what is happening inside a human body. The engineer cannot simply look inside a pipe anymore than a doctor can look inside a blood vessel. Pipes are not transparent and they are usually covered by several feet of earth. The wise engineer must carefully probe the system using gages and meters in much the same manner that a doctor uses manometers and stethoscopes.

The engineer has another problem. While money often is no object where a person's health is concerned, money (or the lack of it) is a serious problem where water distribution systems are concerned. The difference between an adequate engineer and an exceptional one is that the adequate engineer can make a system work, while an exceptional one can do it at the minimum cost. Therefore, this book is intended not only to enable the engineer to understand how distribution systems work, but to enable him to make them work better at the lowest possible cost. (Masculine gender pronouns are used throughout this book. This is is not intended to slight female engineers, but because masculine pronouns are traditionally used to refer to both sexes, and are much easier to use than "he/she" notation.)

The book is intended for the practicing engineer who has had an introductory course in hydraulics or fluid mechanics and still remembers some basic concepts from algebra and just a tiny bit of calculus. Many sections of the book are written in almost cookbook style so the reader can easliy apply the concepts to real problems. Presenting material in this manner is dangerous, as it opens the possibility of an engineer blindly applying the techniques without understanding the concepts and limitations involved. For this reason, the derivation of every equation is presented or referenced so the engineer will understand when it is correct to apply a given technique.

1.2.   OVERVIEW

This book first presents a description of the hydraulics of distribution systems, followed by discussion of what can be referred to as infrastructure analysis (in Chapters 6 and 7). All of these analyses require data about the system. Collection of data is described in Chapter 8. Each of the chapters is described briefly below.

Chapter 2 is a review of the basic hydraulics required to solve water distribution problems. Simply stated: water flows downhill, and flow in equals flow out plus change in storage. Exactly how to calculate these quantities is covered in the chapter.

When a lot of pipes are connected together the principles presented in Chapter 2 still hold, but the mathematics involved with solving the equations become much more difficult. Chapter 3 describes how to set up and solve these problems.

For all but the simplest problems, the engineer will probably want to use a computer model of the system. Chapter 4 describes how to select, calibrate and use these models.

In most cases, hydraulic analyses are carried out to assist in pipe and pump design. Picking the pipe size to minimize life-cycle costs is still something of an art, but some techniques that make this task easier are presented in Chapter 5.

Once a pipe is installed, there is no assurance that it will maintain its carrying capacity indefinitely since tuberculation or scaling is likely to occur. Cleaning and lining the pipe can solve this problem in many cases. Chapter 6 describes how to select the pipes that should be cleaned and lined based on economic criteria.

As pipes age they not only lose their internal carrying capacity, they also become more subject to breakage. At some point in time it becomes economical to replace a pipe or resort to some other remedial measure instead of merely continuing to repair breaks. Chapter 7 describes how to select the pipes that should be repaired or replaced. The common, and often serious, problem of excessive "unaccounted-for" water is also discussed in this chapter, along with loss detection and prevention methods that may be used to ameliorate the problem.

All of the previous chapters contain analytical techniques for addressing water distribution problems. These techniques require some data on the conditions (e.g., flow rates, pressures, pipe roughness) of the system, and the analyses are only as good as the data used. Chapter 8 describes the usual tests for gathering these data.

1.3.   DISTRIBUTION SYSTEM COMPONENTS

There is no such thing as a typical water distribution system. Each one has some unique characteristics due to the water source, service area topography, history of the system, etc. In general, all that can be said is that there are water source(s) and water users and that they are connected by pipes. The pipes can be made of ductile or cast iron, steel, concrete with or without embedded cylinders, various types of plastics, asbestos cement, plus some other innovative materials, and may be connected in an almost limitless number of configurations.

There may be a single source such as a central water treatment plant and pumping station, or water may be supplied by a large number of wells. While pumps are a common component of systems, where the source is at a sufficiently high elevation, the system may not have any pumping.

Most systems contain some storage capacity in the form of tanks which are connected directly to the system (i.e., float on the system), from which water must be pumped or which hold water under pressure (e.g., hydropneumatic tanks).

Valves are required to shut off lines, suppress surges, release air, allow air to enter, drain pipes, control pressure, or simply ease in the operation of other valves. A hydrant is actually a special type of valve that releases water for fire fighting.

Booster pumping may be required to provide adequate pressure in certain portions of a system when there is significant variation in elevation or use rate. On the other hand, pressure reducing valves serving just the opposite purpose may be needed.

Thus, water distribution systems are indeed complicated creatures. However, with the proper design and operation, they can provide excellent service and superb reliability at reasonable cost.

1.4. UNITS

In working with water distribution systems, the engineer must deal with a wide array of measurable quantities, including velocity, flow, pressure, density, viscosity, and length. These quantities are sometimes difficult to work with, not because they are inherently confusing but because they are expressed in so many different kinds of units.

Flow can be expressed in million gallons per day (mgd or mg/day), gallons per minute (gpm or gal/min), cubic meters per second (m^3/sec) or acre-feet per day (ac-ft/day), to name just a few common units. The concept of pressure is even more difficult to handle, since it can be expressed either as pressure or as the height of a liquid column that can be supported by that pressure. Units such as pounds per square inch (psi), feet of water (ft H_2O), millimeters of mercury (mm Hg), kilopascals (kPa), and atmospheres are just some of the ways the water distribution engineer may find pressure expressed.

A sign of the experienced engineer in the water field is the ability to deal with problems in any kind of units. The reader is encouraged to study the list of conversion factors in Appendix A, and become familiar with some of the more commonly used factors. Engineers working with water distribution systems should be able to state "1 psi is equivalent to 2.31 ft of water" and "1 mgd equals 694 gpm" without the need of the tables.

In this book pressure will generally be expressed in pounds per square inch and flow in gallons per minute, although other units are used occasionally to illustrate their use.

The engineer working with water distribution systems should also develop a feel for typical values of pressure, velocity, and head loss encountered in systems. The engineer should know that pressure in a water distribution system usually varies from 60 to 80 psi (400 to 550 kPa); velocity is usually between 1 and 8 ft/sec (0.3 and 3. m/sec) and pipeline head loss rate generally ranges from 1 ft to 10 ft per 1000 ft of pipe (1 to 10 meters per 1000 m). Having a feel for typical values of these terms makes it much easier for the engineer to tackle problems and locate errors in calculations.

REVIEW QUESTIONS

1. Which is larger:
   a. 1 mgd or 1 cubic meter/second
   b. 100 gpm or 1 cfs
   c. 1 atmosphere or 10 psi
   d. 1 kPa or 1 in. water
   e. 100 mm or 1 in.
2. If the static head decreases by 1 psi in 100 ft, would the head loss rate (i.e., slope of hydraulic grade line) be considered large?
3. If pressure at the base of a water tank is 1000 kPa, what is the water level in the tank? Is the pressure typical of pressures found in water distribution systems?

2 | REVIEW OF CLOSED CONDUIT HYDRAULICS

2.1. OVERVIEW

Solving many water distribution system design and operation problems requires an understanding of the equations of closed conduit hydraulics. Usually the solution process involves simultaneous consideration of the energy and continuity equations and some independent relationship describing head loss. In this chapter, these equations are presented and methods for solving them for simple situations are described.

The most important equations are the continuity, momentum, and energy equations. In pipe flow, these equations can be used in their integral form (when average velocity and pressure are required), or in their differential form (when information on such things as velocity distribution within the pipe are required). Most practical problems may be solved by straightforward application of the integral forms, and hence they are emphasized in this chapter. There are, however, some instances where the differential forms are required, so the differential equations are also presented. Readers having somewhat weak backgrounds in differential equations may skip the sections on the differential forms without seriously affecting their ability to use the remainder of the book. However, consideration of the differential forms is very helpful in developing an understanding of the processes occurring in pipe flow, so time spent in reading those sections is worthwhile.

Turbulent flow exists in virtually all situations encountered in water distribution systems. However, since the equations describing laminar flow can be derived analytically, a discussion of them is included in this chapter to illustrate how head loss equations can be developed from momentum considerations.

To solve practical problems the energy equation must be generally coupled with one of several other equations that may be used to predict head loss. The more theoretically correct Darcy–Weisbach equation is presented first, along with a discussion of techniques for calculating the friction factor required to use this equation. The more popular Hazen–Williams equation and other empirical expressions are then presented and their appropriateness for particular situations is discussed.

These equations can be used to predict head loss in straight sections of pipe, and in most cases these are the most important head losses. In some instances, however, "minor" losses which occur at bends and valves can become significant, so methods for predicting them are also presented.

2.2. CONTINUITY EQUATION (INTEGRAL FORM)

The continuity equation is nothing more than the law of conservation of mass passing which simply states that the mass passing into a control volume minus the mass out of the volume is equal to the change in mass stored in that volume. This can be written

$$m_{in} - m_{out} = \Delta m_{store} \qquad (2.2.1)$$

where

$$m_{in} = \text{mass in, } M$$
$$m_{out} = \text{mass out, } M$$
$$\Delta m_{store} = \text{change in storage, } M.$$

Dividing through by density (ρ) (since water can be considered an incompressible fluid in most cases) and a unit of time (t), converts Eq. (2.2.1) into units of volumetric flow ($m/\rho t = Q$). This means that Eq. (2.2.1) can be rewritten

$$Q_{in} - Q_{out} = \Delta S / \Delta t \qquad (2.2.2)$$

where

$$Q_{in} = \text{flow in, } L^3/T$$
$$Q_{out} = \text{flow out, } L^3/T$$
$$\Delta S = \text{change in volume in storage, } L^3$$
$$\Delta t = \text{time period, } T.$$

A more common way of writing Eq. (2.2.2) is to define inflows to be positive, outflows to be negative (since a negative outflow is simply an inflow), and to let t be small. Thus, for n pipes meeting at a point, Eq. (2.2.2) reduces to

$$Q_1 + Q_2 + Q_3 + \cdots + Q_n = dS/dt \qquad (2.2.3)$$

where

$$Q_1 = \text{flow in through pipe 1, } L^3/T$$
$$Q_2 = \text{flow in through pipe 2, } L^3/T$$

$$\vdots$$

$$Q_n = \text{flow in through pipe } n, L^3/T$$
$$dS/dT = \text{time rate of change of storage volume, } L^3/T$$

Eq. (2.2.3) is the continuity equation that will be used for water distribution problems.

EXAMPLE. Consider the system in Fig. 2.1. Determine the rate (in ft/hr) at which the water level in the tank is rising or falling.

Define flow into the tank as positive, and convert all flows into a common unit (say, cubic feet per second) to give

$$Q_1 = (4000 \text{ gpm}) (0.00223 \text{ cfs} / \text{gpm})$$
$$= 8.92 \text{ cfs}$$
$$Q_2 = (-0.20 \text{ } m^3/s) (35.32 \text{ cfs}/m^3/s)$$
$$= -7.06 \text{ cfs}$$
$$Q_3 = (-3.0 \text{ mgd}) (1.547 \text{ cfs}/mgd)$$
$$= -4.64 \text{ cfs.}$$

Summing the flow gives

$$dS/dT = -2.78 \text{ cfs.}$$

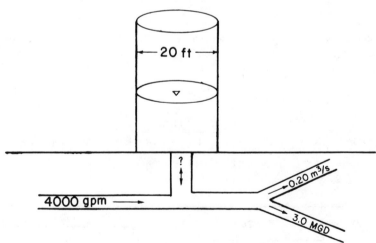

Fig. 2.1. Continuity equation example.

To convert this flow out of the tank to a corresponding rate of change in water surface elevation, determine the cross-sectional area of the tank and divide it into the outflow

$$\text{Area} = \frac{3.14 \; (20 \; \text{ft})^2}{4} = 314 \; \text{ft}^2$$

$$\frac{dh}{dt} = \frac{1}{A}\frac{dS}{dt} = \left(\frac{-2.78 \; \text{cfs}}{314 \; \text{sq. ft}}\right)\left(\frac{3600 \; \text{sec}}{\text{hr}}\right)$$

$$= -31.9 \; \text{ft/hr}.$$

An equation that is often used with (and occasionally mislabeled as) the continuity equation is the following relationship between flow and velocity in a conduit:

$$Q = AV \qquad\qquad (2.2.4)$$

where

Q = flow rate at a given cross section, L^3/T
A = cross-sectional area, L^2
V = average velocity at the cross section, L/T.

Eq. (2.2.4) is often used in pipe flow problems to convert between velocity and flow units and is especially helpful in problems in which pipe diameter changes between cross sections.

EXAMPLE. Given the two pipes in series in Fig. 2.2, find the flow rate in gallons per minute and the velocity at cross section 2 in m/sec.
First find the flow rate using Eq. (2.2.4):

$$Q_1 = A_1 V_1$$

$$= \frac{3.14 \; (0.25 \; \text{m})^2 \; (1 \; \text{m/s})}{4}$$

$$= (0.049 \; \text{m}^3/\text{s}) \; [15845 \; \text{gpm}/(\text{m}^3/\text{s})]$$
$$= 776.3 \; \text{gpm}$$

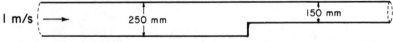

I m/s ⟶   250 mm   150 mm

Fig. 2.2. Continuity equation example for pipes in series.

Using the continuity equation

$$Q_1 = Q_2 = 0.049 \text{ m}^3/\text{sec}.$$

At section 2,

$$V_2 = \frac{0.049 \text{ m}^3/\text{sec}}{3.14(0.15 \text{ m})^2/4} = 2.77 \text{ m/sec}$$

2.3. CONTINUITY EQUATION (DIFFERENTIAL FORM)

In the previous section, the integral form of the continuity equation which is used for macroscopic flow calculations was presented. This expression is applicable when it is sufficient to know only average velocities. Sometimes it is desirable to know not only the average velocity in the pipe but how the velocity is distributed across a section of pipe. In such cases the continuity equation must be written in differential form, which for an incompressible fluid is

$$(\nabla \cdot \mathbf{v}) = 0 \tag{2.3.1}$$

where

$$\nabla = \text{differential operator}$$
$$\mathbf{v} = \text{velocity vector.}$$

By itself this equation is not especially useful, but it can be solved in conjunction with other differential equations describing fluid motion to predict velocity distribution. To evaluate the ∇ operator, it is necessary to know the coordinate system being used. For pipe flow problems, cylindrical coordinates, as show in Figure 2.3, are usually used. Thus, Eq. (2.3.1) can be written

$$\frac{1}{r}\frac{\partial}{\partial r}(rv_r) + \frac{1}{r}\frac{\partial v_\theta}{\partial \theta} + \frac{\partial v_z}{\partial z} = 0 \tag{2.3.2}$$

where

$$r = r \text{ coordinate, } L$$
$$v_r = \text{velocity in } r \text{ direction, } L/T$$
$$v_\theta = \text{velocity in } \theta \text{ direction, } L/T$$
$$v_z = \text{velocity in the } z \text{ direction, } L/T.$$

Note that Eqs. (2.3.1) and (2.3.2) pertain to one point in the fluid, while the

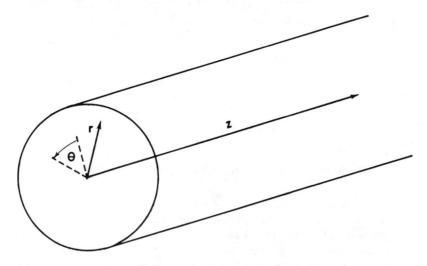

Fig. 2.3. Definition sketch for cylindrical coordinates.

integral form of the continuity equation pertains to spatially averaged flow. Usually v_z is the component of the velocity vector of interest.

The average velocity can be determined from the velocity distribution, by integrating the velocity distribution over the area of the pipe as

$$V = \frac{\int_A v_z(r,\theta)\, dA}{A} \tag{2.3.3}$$

where A = area of the pipe, L^2.

Equation (2.3.3) refers to a straight section of pipe in which the velocity is not changing in the downstream direction. If the pipe is circular and there is no circulation in the pipe, Eq. (2.3.4) becomes

$$V = \frac{\int_0^{2\pi} \int_0^R v_z(r)r\, dr\, d\theta}{A} \tag{2.3.4}$$

where R = radius of the pipe, L. Use of this equation will be demonstrated later.

For many problems the velocity distribution is not known as a continuous function of r but rather as discrete observations of velocity at various points in the pipe. When the velocity is known only at discrete points, Eq. (2.3.4) can be approximated by

$$V = \frac{\sum\limits_{i=1}^{n} v_i A_i}{A} \tag{2.3.5}$$

where

n = number of observations
v_i = velocity at ith observation, L/T
A_i = annular area corresponding to ith velocity, L^2.

This formula may be used to determine average velocity from the results of a traverse of the pipe using a pitot rod. It is convenient to divide the total cross-sectional area of the pipe into annular areas because v usually varies with r rather than θ or z.

2.4. MOMENTUM EQUATION (INTEGRAL FORM)

The momentum equation follows directly from Newton's second law of motion. That is,

$$F = ma = \frac{d(mv)}{dt} \tag{2.4.1}$$

where

F = component of force vector, ML/T^2
m = mass, M
a = acceleration, L/T^2
v = component of velocity vector, L/T
t = time, T.

Any difference in momentum between two points in a pipe (with no inflow or outflow between the two points) must be caused by a force acting on the fluid. If the force acts over time (Δt), Eq. (2.4.1) can be reorganized to give

$$F = \frac{\Delta mv}{\Delta t} = \frac{m_2 v_2}{\Delta t} - \frac{m_1 v_1}{\Delta t} \tag{2.4.2}$$

where the subscripts 1 and 2 denote the upstream and downstream points, respectively.

The mass flow rate is $m/\Delta t$ and can be replaced by the product of the density (ρ) and the volumetric flow rate (Q) to give

$$F = \rho_2 Q_2 v_2 - \rho_1 Q_1 v_1 \tag{2.4.3}$$

where

$$\rho = \text{density}, \; M/L^3$$
$$Q = \text{flow rate}, \; L^3/T.$$

Since water is incompressible ($\rho_1 = \rho_2$) and there are no inflows ($Q = Q_1 = Q_2$), Eq. (2.4.3) reduces to

$$F = \rho Q \, (v_2 - v_1). \tag{2.4.4}$$

There are actually three separate equations like Eq. (2.4.4), since both F and v are vectors and, hence, may have components in three dimensions.

Equation (2.4.4) represents the force resulting in a change in momentum along a single streamline, but is generally applied to a pipe in which velocity varies over the cross section. In such a situation, one should actually use

$$\rho \int_A v^2 \, dA \tag{2.4.5}$$

as the momentum at a cross section. To simplify the algebra, it is possible to introduce a correction factor (β) such that

$$\beta \, V^2 A = \int_A v^2 \, dA \tag{2.4.6}$$

where V = average velocity, L/T. β = momentum correction factor.

Using Eq. (2.4.6), it is only necessary to know the average velocity to solve problems involving momentum in a pipe. β is dimensionless and is slightly greater than unity for turbulent flow. In most cases, however, it can be considered to be equal to unity with neglible error.

In water distribution system analysis, the momentum equation is most often used to compute forces acting on pipe bends and contractions so that adequate constraints can be designed. To determine the net force acting on a bend or contraction, the momentum force must be added to the pressure force. In most cases the pressure forces are considerably larger than forces resulting from changes of momentum.

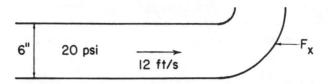

Fig. 2.4. Example of forces acting on bend.

EXAMPLE. Consider the bend shown in Fig. 2.4. Determine the force in the horizontal direction that must be exerted to hold the pipe in place.

It is first necessary to determine the pipe cross-sectional area and flow as follows:

$$A = 28.3 \text{ sq. in.} = 0.196 \text{ sq. ft}$$

$$Q = AV = (0.196 \text{ sq. ft})(12 \text{ ft/sec})$$

$$= 2.36 \text{ cfs.}$$

The net horizontal force on the bend is

$$F_x = pA - pQ(V_2 - V_1).$$

The horizontal component of the flow velocity downstream from the bend is zero ($V_2 = 0$), so

$$F_x = (20 \text{ psi})(28.3 \text{ sq. in.}) - (1.93 \text{ slug/ft}^3)(2.36 \text{ cfs})(-12 \text{ ft/sec})$$

$$= 556 + 55$$

$$= 621 \text{ lb}_f.$$

2.5. MOMENTUM EQUATION (DIFFERENTIAL FORM)

The differential form of the momentum equation as presented below is a somewhat imposing looking expression which accounts for all the forces acting to change the momentum of a tiny volume of water. To avoid being overwhelmed by the equation, the reader need only remember that it is simply a very precise way of saying that

$$ma = F \tag{2.5.1}$$

while including the effects of all the forces and accelerations a flowing fluid in a water distribution system normally experiences.

The differential form of the momentum equation is known as the Navier–Stokes equation and is one of the most important equations in fluid mechanics. Since it is a vector expression, there are actually three separate components, one for each of the three dimensions in pipe flow (r, θ, and z). Since in pipe flow problems the most important component of velocity is in the z direction, the z-momentum equation for a Newtonian fluid with constant density and viscosity is given below:

$$\rho \left(\underset{A}{\frac{\partial v_z}{\partial t}} + \underset{B}{v_r \frac{\partial v_z}{\partial r}} + \underset{C}{\frac{v_\theta}{r} \frac{\partial v_z}{\partial \theta}} + \underset{D}{v_z \frac{\partial v_z}{\partial z}} \right) = \underset{E}{-\frac{\partial p}{\partial z}}$$

$$+ \underset{F}{\mu \left[\frac{1}{r} \frac{\partial}{\partial r} \left(r \frac{\partial v_z}{\partial r} \right) \right.} + \underset{G}{\frac{1}{r^2} \frac{\partial^2 v_z}{\partial \theta^2}} + \left. \underset{H}{\frac{\partial^2 v_z}{\partial z^2}} \right] + \underset{I}{\rho g_z} \qquad (2.5.2)$$

where

v_r, v_θ, v_z = velocity components in the r, θ, and z directions respectively, L/T
ρ = density, M/L^3
μ = viscosity, M/LT
p = pressure, M/LT^2
g_z = gravitational acceleration in the z direction, L/T^2.

While Eq. (2.5.2) appears extremely complicated and cannot be solved analytically, some insights into the forces that move fluids can be gained by examining the individual terms in the equation. For a fairly readable derivation of this equation, the reader is referred to *Transport Phenomena* (Bird, Stewart and Lightfoot, 1960). This is done below first for laminar flow, then the adjustments required for turbulent flow are described. (Note that each term has units of force per unit volume.)

The A term refers to the change in momentum when the velocity changes with time (for example when a pump starts or a tank drains quickly). The B term refers to acceleration caused by flow in the radial direction, and can be important in problems involving porous pipe such as diffusers or well screens. The C term refers to acceleration in the θ direction and is important when the flow is moving along a spiral path. For pipe flow, the D term is usually the most important of the acceleration terms on the left-hand side of the equation since it is important at expansions and contractions in the pipe.

While the left side of the equation describes the accelerations caused by forces acting on the fluid, the right side describes the forces themselves. Three types of forces are included: pressure, viscous, and gravitational. The E term refers to the pressure gradient and is the primary driving force in pipe flow. The F, G, and H terms are viscous drag terms. Since there is usually very little change in velocity gradient in the θ and z directions, the F term is usually the most important. The I term refers to gravitational forces and is largest for vertically mounted pipes and zero for horizontally mounted pipes. g_z can be determined as $g \cos \alpha$, where g is acceleration due to gravity and α is the angle between z and the vertical direction.

If one divides through Eq. (2.5.2) by characteristic length, velocity, and density values, a dimensionless equation is obtained. The expression in front of the viscous force term is $\mu / LV\rho$ (where μ, L, V, and ρ are the characteristic viscosity, length, velocity, and density, respectively) which is simply the reciprical of the Reynolds number (N_r). When the Reynolds number is small, viscous forces are large in comparison with other forces, while if the Reynolds number is large, viscous forces are relatively unimportant. This is why many coefficients in hydraulics are correlated with the Reynolds number.

EXAMPLE. Solve the z-component momentum equation to determine the velocity profile for steady, laminar flow in a straight, horizontal, circular pipe of radius R with constant pressure gradient P_z and no inflow or outflow.

To solve this problem, or any problem involving Eq. (2.5.2), it is first necessary to eliminate terms which are zero or very small so that the mathematics can be made easier. Since the flow is steadv. $\partial v_z / \partial t = 0$. Since the flow is laminar in a straight, circular pipe, there is no radial or rotational velocity component, so that $v_r = v_\theta = 0$. Since the pipe is horizontal, $g_z = 0$. Since the pipe is of constant radius and there are no inflows or outflows, there will be no change in velocity in the z direction and $\partial v_z / \partial z = \partial v_z^2 / \partial z^2 = 0$.

With these simplifications, Eq. (2.5.2) can be reduced to

$$P_z = \frac{\mu}{r} \frac{d}{dr}\left(r \frac{dv_z}{dr} \right).$$

Multiplying through by $(r/\mu)\, dr$ and integrating yields

$$\frac{P_x r^2}{2\mu} = r \frac{dv_z}{dr} + C_1$$

To determine the constant of integration, note that the velocity gradient must be finite at $r = 0$ (the velocity gradient can be shown to be zero). For the velocity gradient to be finite, the constant of integration must be zero. This is necessary if there are to be no discontinuities in the velocity profile. Integrating again yields

$$v_z = \frac{P_x r^2}{4\mu} + C_2.$$

Since the fluid does not slip at the wall, the boundary condition, $v_z = 0$ at $r = R$, can be used to determine the constant of integration. Substituting for C_2 and solving for velocity yields

$$v_z = \frac{-P_x R^2}{4\mu}\left[1 - \left(\frac{r}{R}\right)^2\right].$$

This equation indicates that the velocity profile for steady, laminar, flow in a horizontal, circular pipe can be described by a parabola. This expression can be substituted into Eq. (2.3.4) and integrated to determine the average velocity in the pipe as

$$V = \frac{P_z R^2}{8\mu}.$$

Multiplying by the cross-sectional area gives

$$Q = AV = \frac{-\pi P_x R^4}{8\mu}$$

which is the Hagen–Poiseuille equation for laminar flow in circular tubes.

While the above equations describe laminar flow in pipes, virtually all flow in water distribution systems is turbulent. In turbulent flow, the velocity at any point at any time can be separated into its temporally averaged velocity and a fluctuation from this average, as follows:

$$v = \bar{v} + v' \qquad (2.5.3)$$

where

$$v = \text{velocity, } L/T$$
$$\bar{v} = \text{temporally averaged velocity, } L/T$$
$$v' = \text{fluctuation from } \bar{v}, L/T.$$

Substituting Eq. (2.5.3) for every term in Eq. (2.5.2) would yield an extremely complicated equation which could not be solved for even the simplest boundary conditions. Fortunately for most engineering problems in which velocity distributions must be known it is only necessary to know the temporally averaged velocity. This means that Eq. (2.5.2) can be written for turbulent flow as

$$\rho\left(\frac{\partial \bar{v}_z}{\partial t} + \bar{v}_r\frac{\partial \bar{v}_z}{\partial r} + \frac{\bar{v}_\theta \partial \bar{v}_z}{r\ \partial \theta} + \bar{v}_z\frac{\partial \bar{v}_z}{\partial z}\right) = -\frac{\partial p}{\partial z}$$
$$+ (\mu + \epsilon)\left[r\frac{\partial}{\partial r}\left(r\frac{\partial \bar{v}_z}{\partial r}\right) + \frac{1}{r^2}\frac{\partial^2 \bar{v}_z}{\partial \theta^2} + \frac{\partial^2 \bar{v}_z}{\partial z^2}\right] + \rho g_z \quad (2.5.4)$$

where $\epsilon = $ eddy viscosity, M/LT.

Equation (2.5.4) is referred to as the Reynolds equation and looks much like Eq. (2.5.2) except that the velocities have been replaced by time-averaged velocities. There is also one additional variable in the viscous force term, the eddy viscosity. This eddy viscosity (not to be confused with the kinematic viscosity $\nu = \mu/\rho$) is considerably more difficult to work with than the absolute viscosity, which is a property of the fluid, because the eddy viscosity is a property of the flow.

The addition of the eddy viscosity to the equations describing turbulent flow is significant in that it accounts for the different ways momentum is transported in fluids in laminar and turbulent flow. In laminar flow momentum is transported from the high-velocity areas in the center of the pipe out toward the wall by molecular drag between the molecules. In turbulent flow, this process still occurs but it is much less effective than the transport of momentum through eddies (i.e., $\epsilon \gg \mu$). Consequently, the velocity profile in a pipe with turbulent flow tends to be much flatter than that in laminar flow. The shape of velocity profiles in pipes will be described in more detail in Section 2.7.

2.6. ENERGY EQUATION

Everyone knows that water flows downhill. It is the energy equation, coupled with some expression for head loss, that enables the engineer to determine which direction is hydraulically downhill and how fast the water will flow. The energy equation states that, given the energy at point 1, the energy at point 2 equals the energy at 1, plus any work done on the fluid by pumps, minus any work done by the water on turbines, minus any energy losses due to friction. Mathematically this can be stated as follows:

$$E_2 = E_1 + W - H \qquad (2.6.1)$$

where

$$E_2 = \text{energy at point 2, } ML^2/T^2$$
$$E_1 = \text{energy at point 1, } ML^2/T^2$$
$$W = \text{net work done on the fluid, } ML^2/T^2$$
$$H = \text{friction energy loss, } ML^2/T^2.$$

The energy in the fluid exists in three forms: kinetic energy, potential energy due to elevation, and internal energy (pressure). The total energy may be expressed as

$$E = (mv^2/2) + mgz + pm/\rho \qquad (2.6.2)$$

where

m = mass of the fluid, M
v = velocity, L/T
g = acceleration due to gravity, L/T^2
z = elevation, L
p = pressure, M/LT^2
ρ = density, M/L^3.

Each of the three terms on the right-hand side of Eq. (2.6.2) represents one of three forms of energy:

1. The term $mv^2/2$ refers to the kinetic energy. It is generally most important for problems in which the pipe diameters change; in other situations the kinetic energy remains roughly constant. In water distribution system problems, the kinetic energy term is usually small in comparison to the other terms.

2. The mgz term refers to the energy the fluid has because of its position in the gravitational field. It is important to remember to use the same datum for all elevations in a given problem.

3. The pm/ρ term refers to the amount of energy stored in the form of pressure.

It is customary to divide Eq. (2.6.2) by gm. This allows all the energy terms to be expressed in units of length (equivalent to energy per unit weight). Making this conversion and substituting Eq. (2.6.2) into (2.6.1), noting $\gamma = \rho g$, gives

$$\frac{v_1^2 - v_2^2}{2g} + z_1 - z_2 + \frac{P_1 - P_2}{\gamma} = -w + h \qquad (2.6.3)$$

where

w = net work done on fluid, L
h = friction energy loss, L
γ = specific weight, $M/L^2 T^2$.

All the variables in Eq. (2.6.3) can be positive or negative except for the h term, which can only be positive. Energy lost to friction cannot be recovered, so it is this term that indicates the direction of flow. Since flow was from 1 to 2 above, then h has to be positive. If the value determined for h turned out to be negative, it would indicate that flow was actually from 2 to 1.

Each of the terms in Eq. (2.6.3) has a special name, given below:

$$\frac{v^2}{2g} = \text{velocity head, } L$$

$$\frac{P}{\gamma} = \text{pressure head, } L$$

z = elevation head, L

w = lift (when referring to pumps), L

h = friction head, L.

The sum of the velocity, pressure and elevation heads is called the *energy head* and is also referred to as the *energy grade line elevation*. The slope of this line is called the *energy gradient*. The sum of the elevation head and pressure head is called the *static head* or *piezometric head,* and its elevation is called the *hydraulic grade line elevation*. The slope of the hydraulic grade line is called the *hydraulic gradient.*

Water flows from areas of high head to low head. In most water distribution problems the difference between the hydraulic grade line and energy grade line elevations are small, since the velocity is usually fairly low in water distribution systems. Pumps (or turbines) introduce abrupt breaks in the energy grade line where fairly large amounts of energy are added to (or extracted from) the fluid.

As written in Eq. (2.6.3), the energy equation only holds along individual streamlines in the fluid. For practical problems it is important to be able to apply it to the pipe as a whole. To do this it is necessary to develop a correction factor relating the average kinetic energy to the kinetic energy calculated based on average velocity. It is known that the kinetic energy per unit mass flowing past a cross section of pipe is a product of the mass flow rate $\int_A v \, da$ and the velocity squared (v^2). This kinetic energy term may be set equal to the kinetic energy based on the average flow rate V multiplied by a correction factor α to give

$$KE = \int_A \rho v^3 \, dA = \alpha \rho \int V^3 \, dA \qquad (2.6.4)$$

where

α = kinetic energy correction factor
v = velocity, L/T
V = average velocity, L/T.

Solving for α for an incompressible fluid yields

$$\alpha = \frac{1}{A} \int_A \left(\frac{v}{V}\right)^3 dA. \tag{2.6.5}$$

For turbulent flow problems the kinetic energy correcton factor is approximately 1.1. However, since the velocity head in water distribution system piping is small, this term generally can be ignored with negligible error. The velocity head terms in equation (2.6.3) in integral form should therefore be written as

$$\frac{\alpha(V_1^2 - V_2^2)}{2g}$$

EXAMPLE. The pressure in a 6-in-diameter line carrying 5 mgd is 60 psi at elevation 100 ft. What is the pressure at the downstream end of the pipe if the elevation is 90 ft, the pipe is 1000 ft long, the head loss is 20 ft, and there are no pumps along the pipeline? Since there is no change in velocity and no pumping, the energy equation can be simplified to

$$z_1 - z_2 + \frac{P_1 - P_2}{\gamma} = h.$$

Solving for P_2 yields

$$\begin{aligned} P_2 &= P_1 + (z_1 - z_2 - h)(\gamma/144) \\ &= 60 + (100 - 90 - 20)(62.4/144) \\ &= 55.7 \text{ psi.} \end{aligned}$$

2.7. VELOCITY DISTRIBUTIONS

In solving most water distribution system problems, it is only necessary to know the average velocity of the water flowing through the pipes. In some cases, however, it is necessary to be able to describe the manner in which velocity is distributed in the pipe or to calculate average velocity given the velocities at several points. In Section 2.5, the differential form of the momentum equation was solved to give the velocity distribution for laminar flow. This equation is fairly easy to solve for laminar flow because momentum transport in water resulted from the effects of viscosity and the relationship between shear stress and velocity gradient could be given by Newton's Law of Viscosity:

$$\tau_{yx} = -\mu \frac{dv_x}{dy} \tag{2.7.1}$$

where

τ_{yx} = shear stress in the x direction caused by a velocity gradient in the y direction, M/LT^2.

To better understand this mechanism, it is helpful to interpret the τ_{yx} term in Eq. (2.7.1) as the flux of x momentum in the y-direction. Momentum is carried from areas of higher velocity in the center of the pipe to areas of lower velocity near the wall. This is analogous to Fourier's Law of Heat Transfer (heat flow = conductivity times temperature gradient) and Fick's Law of Diffusion (mass flux = diffusivity times concentration gradient). That is, momentum flows from areas of high velocity to low velocity just as heat flows from hot areas to cold areas and chemical species diffuse from areas of high concentration to areas of low concentration.

However, as previously discussed, eddies are very important for transporting momentum in turbulent flow. Since eddies are much more effective than the mechanism described by Newton's Law of Viscosity, velocity profiles in laminar and turbulent flow tend to be significantly different.

To determine the velocity profile in turbulent flow, it is necessary to solve the Reynolds equation which contains the eddy viscosity. Because of the difficulty in analytically describing the eddy viscosity, formulas describing the velocity profile in turbulent flow involve some empiricism. Several such empirical relationships are available for specific applications, depending on the roughness of the pipe. A comparison of velocity profiles in laminar and turbulent flow is shown in Figure 2.5.

Before considering the velocity profile equations, in detail, it is important to understand the nature of turbulent flow in pipes. There are actually two distinct flow regions in the pipe: a thin layer along the wall, called the *boundary layer,* in which viscous forces predominate, and the main body of the flow which is referred to as the *turbulent core.* The shape of the velocity profile is highly dependent on the shear stress at the wall (τ_0). This wall shear stress is usually expressed in velocity units as shown below, and is referred to as the friction velocity (v^*):

$$v^* = \sqrt{\tau_0/\rho} \qquad (2.7.2)$$

where

v^* = friction velocity, L/T
τ_0 = wall shear stress, M/LT^2
ρ = density, M/L^3.

The friction velocity does not necessarily correspond to the actual velocity at any point in the pipe, but is rather an indication of pipe roughness. The ratio

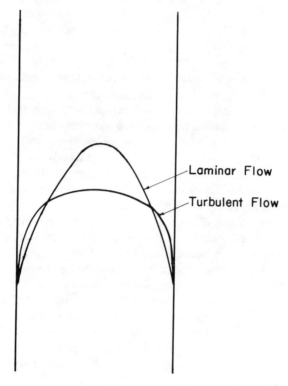

Fig. 2.5. Velocity profiles in laminar and turbulent flow.

of the friction velocity to the average velocity can be related to a parameter f, which is called the *friction factor*, by the formula

$$v*/V = \sqrt{f/8} \qquad (2.7.3)$$

where

$$V = \text{average velocity, } L/T$$
$$f = \text{friction factor.}$$

There are essentially two types of formula for determining velocity profiles in turbulent flow: (1) smooth flow equations, for cases in which the roughness of the wall is so smooth there is a boundary layer in which the flow is laminar, and (2) rough flow equations in which the height of the roughness elements is so great that a laminar boundary layer cannot exist. These formulas can be developed by plotting some form of dimensionless velocity versus a dimensionless distance from the wall on semi-log or log-log graph paper. (In some cases, the expression *smooth pipe* is used to describe a pipe in which flow is almost always hydraulically smooth, while *rough pipe* is used to describe pipes in which flow is usually hydraulically rough. For the purpose of describing the velocity distribution or head loss characteristics, it is the flow and not the pipe that is smooth or rough.) A few of the more commonly used equations for velocity profile are summarized below.

The Law of the Wall can be developed by plotting dimensionless velocity $(v/v*)$ versus dimensionless distance from the wall $(yv*\rho/\mu)$ on a semi-log graph to yield the following relationship:

$$v/v* = 2.44 \ln (yv*\rho/\mu) + 4.9 \tag{2.7.4}$$

where y = distance from the wall, L. This formula is appropriate outside the laminar boundary layer $(y > 10\mu/v*\rho)$.

The Velocity Defect (or Deficiency) Law is based on the fact that momentum is transported to areas of low velocity at a rate related to the difference between the peak velocity and the velocity at the point of interest. The dimensionless velocity defect is given by $(v_m - v)/v*$ and the Velocity Defect Law is given as

$$\frac{v}{v*} = \left(\frac{v_m}{v*}\right) - \ln\left(\frac{R}{y}\right) \tag{2.7.5}$$

where

$$v = \text{velocity of any point, } L/T$$
$$v_m = \text{maximum velocity, } L/T$$
$$R = \text{pipe radius, } L.$$

The maximum velocity varies from 1.1 to 1.25 times the average velocity in a straight pipe, and can be approximated as follows using the friction factor:

$$v_m = V(1.43 \sqrt{f} + 1). \tag{2.7.6}$$

The Power Law for Velocity can be derived by noting that for smooth flow, the friction factor is proportional to the Reynolds Number to a power. The most general form of the equation is

$$v = v_m(y/R)^{m/(2-m)}. \tag{2.7.8}$$

For Reynolds Numbers less than 100,000, $m = 0.25$ and Eq. (2.7.8) can be written

$$v = v_m(y/R)^{1/7} \tag{2.7.9}$$

which is called the *one-seventh power law*. The exponent in the power law decreases for Reynolds Numbers greater than 100,000. The one-seventh power law may also be written in terms of friction velocity instead of maximum velocity:

$$v = 8.7 \ v*(v*y/v)^{1/7} \tag{2.7.10}$$

where v = kinematic viscosity, L^2/T.

For rough pipe, the viscosity of the fluid becomes less important than the height of the roughness element so the location in the pipe is measured not as distance beyond the laminar boundary layer but rather as distance beyond the roughness elements, as follows:

$$\frac{v}{v*} = 8.5 + 2.5 \ln \left(\frac{y}{\epsilon}\right). \tag{2.7.11}$$

For more information on velocity distributions in pipes, the reader is referred to *Boundary Layer Theory* by Schlichting (1979).

2.8. CALCULATING HEAD LOSS

The remainder of this chapter is devoted to methods for determining the head loss h in the energy equation. Head losses arise because of friction between the fluid and the pipe wall and turbulence within the fluid. It would be convenient if head loss were constant (e.g., 10 ft loss/1000 ft length) or a simple function of velocity or velocity head. Unfortunately, the rate of head loss in a pipe depends not only on velocity and pipe diameter, but also on roughness element size. To further complicate the matter, the relationship between these variables depends on whether the pipe can be considered hydraulically smooth, rough, or somewhere in between. In water distribution systems, head losses also result from bends, valves, and changes in pipe diameter, so that determining h in the energy equation is often more difficult than determining all of the other terms.

In the sections presented below, various formulas for determining h are presented. First, head loss in a straight section of pipe is discussed in a generalized manner. Then, specific formulas for head loss in laminar flow, and turbulent flow in smooth, rough, and transition flow are presented. Each of the turbulent flow head loss equations requires a coefficient. The methods for calculating these coefficients are described and compared. Finally minor losses through bends, valves, and changes in diameter are discussed.

Predicting the head loss for steady flow through a straight pipe involves a balance between the pressure force driving the flow and the drag forces acting at the wall to restrain flow. If the drag forces along the wall do not exactly offset the driving forces due to pressure, then the fluid will be accelerating or decelerating and the flow will be unsteady by definition. The balance between pressure and wall shear forces can be written as follows:

$$\tau_0 A_w = \Delta p A_r \tag{2.8.1}$$

where

τ_0 = shear stress at the wall, M/LT^2
A_w = area of wall over which shear acts, L^2
    = πDL
D = diameter of pipe, L
L = length over which head loss occurs, L
Δp = pressure drop in length L, M/LT^2
A_r = cross-sectional area, L^2
    = $D^2/4$.

Equation (2.8.1) can be reduced to

$$\Delta p = 4L\tau_0/D. \tag{2.8.2}$$

As given in Eq. (2.7.2) and (2.7.3), the wall shear stress can be related to the fluid density and friction factor as follows:

$$\tau_0 = f\rho V^2/8 \tag{2.8.3}$$

where

V = average velocity, L/T
f = friction factor.

Substituting Eq. (2.8.3) into Eq. (2.8.2) and dividing by the specific weight of the fluid yields

$$h = \frac{\Delta p}{\gamma} = f\frac{L}{D}\frac{V^2}{2g} \tag{2.8.4}$$

where γ = specific weight of water, M/L^2T^2. This equation is called the Darcy–Weisbach equation. The term h/L is the slope of the hydraulic and energy grade lines for a pipe of constant diameter, and is usually represented by S. The parameters D, L, and V can all be determined fairly easily for most situations. Unfortunately, f cannot be measured directly and it not only is not a constant, it is not even a constant for a given pipe. Much of the remainder of this chapter describes how to determine f and other parameters which can be used in place of f.

2.9. HEAD LOSS IN LAMINAR FLOW

In laminar flow, it is possible to use the Hagen–Poiseuille equation developed in Section 2.5 as a starting point to develop an analytical expression for f. First, note that the term P_x is actually γS and $Q = V\pi D^2/4$. Substituting these expressions for P_x and Q in the Hagen–Poiseuille equation, and solving for S (noting that $\gamma/\rho = g$) gives

$$S = \frac{32V\mu}{D^2\rho g}. \qquad (2.9.1)$$

Substituting for S in the Darcy–Weisbach equation and solving for f yields

$$f = \frac{64\mu}{\rho VD}. \qquad (2.9.2)$$

Insightful readers will note that $\rho DV/\mu$ is the Reynolds number. Therefore, for laminar flow, the friction factor depends only on the Reynolds number. That is,

$$f = 64/N_r \qquad (2.9.3)$$

where N_r = Reynolds number. This relationship is valid only for the small Reynolds numbers associated with laminar flow (i.e., $N_r < 1000$).

EXAMPLE. Find the friction factor and head loss in 10 m of 10-mm-diameter tubing carrying water ($\nu = 0.005$ cm^2/sec) at 0.02 m/s.
  First determine the Reynolds number

$$N_r = DV/\nu$$

$$= (1\,\text{cm})\,(2\,\text{cm/sec})/(0.005\,\text{cm}^2/\text{sec})$$

$$= 400$$

Equation (2.9.3) gives $f = 64/400 = 0.16$.

Next the head loss can be calculated using the Darcy–Weisbach equation:

$$h = 0.16\left(\frac{10\ \text{m}}{0.01\ \text{m}}\right)\left(\frac{(0.02\ \text{m/sec})^2}{2(9.8\ \text{m/sec}^2)}\right)$$

$$= 0.0033\ \text{m}.$$

2.10. HEAD LOSS IN TURBULENT FLOW

In turbulent flow, the friction factor is a function of both Reynolds number and pipe roughness. The friction factor can be regarded as an interphase momentum transport constant indicating how effective the pipe wall is in absorbing the momentum of the flow. It is critical, therefore, to know the type of flow occurring in the boundary layer in order in calculate f.

As stated previously, there may be a laminar layer along the boundary if the pipe is sufficiently smooth and the flow is sufficiently slow. In such a situation, roughness elements in the pipe wall do not protrude through the laminar boundary layer, and the friction factor depends only on Reynolds number. If the roughness elements begin to affect the friction factor (either the velocity increases or the elements are larger), then the friction factor depends on both Reynolds number and relative roughness (ratio of roughness height to pipe diameter). Eventually as the roughness size or the velocity increases, the laminar boundary layer ceases to exist and the flow is wholly rough. The friction factor in this case depends on the relative roughness. Note that it is the flow and not the pipe that is described as smooth or rough. In fact it possible for a given pipe to have smooth flow at a very low velocity and rough flow at a very high velocity. For example, a 1-ft-diameter pipe with a 0.0001 ft roughness height will have smooth flow at a velocity of 1 ft/sec, rough flow at a velocity of 20 ft/sec, and transition flow at intermediate velocities. For the velocities and diameters found in most water distribution systems, transition flow predominates, although rough flow can be found in older pipes. Smooth flow is the exceptional case.

The distinction between smooth and rough flow is important because it is possible to use fairly simple formulas to calculate f for either case as long as the type of flow that is occurring is known. These are given in the following sections. Where graphical determination of the friction factor is acceptable, it is possible to use a Moody diagram, as shown in Fig. 2.6. This figure gives the friction factor over a wide range of Reynolds numbers for laminar flow and smooth, transition, and rough turbulent flow. Laminar flow is represented as a single line described by the equation $f = 64/N_r$. Turbulent flow is represented by the family of lines in the higher Reynolds numbers. Each line corresponds to a different pipe roughness. The bottom line corresponds to smooth flow. At a low enough Reynolds number, virtually all pipes exhibit smooth flow, but at very high Reynolds numbers only pipes with very low relative roughnesses have lines close to the smooth pipe line. Note also that at high Reynolds numbers the curves pertaining to rougher pipes become flat (i.e., depend only on the roughness and not on Reynolds number).

The quantities shown on the Moody diagram are dimensionless so they can be used with any system of units. The top horizontal axis is for use only with English units for situations when the effect of temperature on the Reynolds number may be ignored.

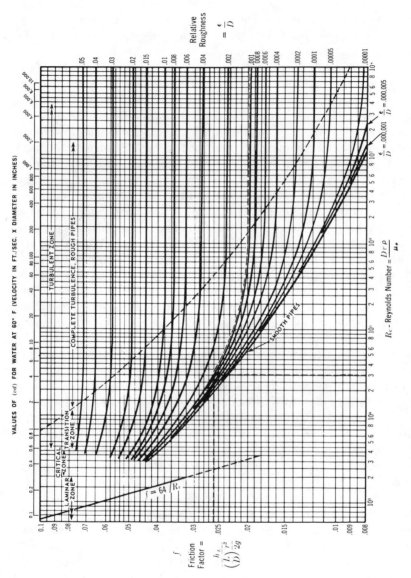

Fig. 2.6. Moody diagram. [Reprinted from Crane Technical Report 40, with permission of The American Society of Mechanical Engineers and Crane Co., which extracted the diagram from an article by L. F. Moody in *Transactions of ASME*, Vol. 66, 1944.]

In addition to the height of the roughness elements, their shape and spacing are also important. The roughness height is actually based on the diameter of sand grains that would give the same head loss. Therefore the friction factor shown on the Moody diagram is only an approximation of the actual friction factor if the roughness elements are irregularly shaped or spaced. Morris and Wiggert (1972) give methods for calculating friction factors for isolated roughness elements, skimming flow over depressions in pipe walls, corrugated and sharp strip roughness elements and spot roughness.

2.11. FRICTION FACTOR FOR SMOOTH FLOW

Examination of the Moody Diagram for smooth flow and modest Reynolds numbers shows that Reynolds number and friction factor can be related by a straight line on a log-log graph. This means the relationship can be represented by an equation of the form $f = aN_r^b$. The coefficients that best fit the data are $a = 0.316$ and $b = -0.25$ which gives

$$f = 0.316/N_r^{0.25}. \qquad (2.11.1)$$

This formula, known as the Blasius equation, is only applicable for Reynolds numbers less than 100,000 since the "straight line" begins to curve for higher Reynolds numbers. An equation that fits the data for a wider range of Reynolds numbers is the von Karman–Nikuradse smooth pipe equation

$$f = [2 \log (N_r \sqrt{f}) - 0.80]^{-2}. \qquad (2.11.2)$$

(Note that in this book log refers to base 10 logarithm, while ln refers to base e logarithms). Unlike the Blasius equation, Eq. (2.11.2) cannot be solved directly for f, so a trial-and-error solution is necessary.

Thus far it has been implied that the expressions for determining f are simply empirical regression equations. Actually, they can be derived directly from the velocity distribution in the pipe. To illustrate this, the Blasius equation will be derived from the one-seventh power law.

First, it is necessary to integrate the velocity profile to develop a relationship between v^* and V:

$$V = \frac{2}{R^2} \int_0^R 8.7 v^* \left[\frac{v^*(R-r)}{\nu} \right]^{1/7} r \, dr. \qquad (2.11.3)$$

Performing the integration and solving for v^* yields

$$v^* = \left(\frac{\nu^{1/7} v}{7.1 R^{1/7}} \right)^{7/8}. \qquad (2.11.4)$$

Substituting Eq. (2.11.4) into Eq. (2.7.3) and solving for f (noting $D = 2R$) gives

$$f = 8 \left[\frac{0.18 \, \nu^{1/8}}{0.917(VD)^{1/8}} \right]^2. \qquad (2.11.5)$$

Since $N_r = VD/\nu$,

$$f = 0.315/N_r^{0.25}$$

which is the Blasius equation.

2.12. FRICTION FACTOR FOR ROUGH FLOW

In fully rough flow there is no laminar boundary layer, so the friction factor depends on the relative roughness. Starting with the rough pipe velocity distribution as given in Eq. (2.7.11), and using a procedure similar to that used to develop the Blasius equation, the following relationship for friction factor can be developed:

$$f = [1.14 + 2 \log (D/\epsilon)]^{-2}. \qquad (2.12.1)$$

Eq. (2.12.1) describes the portion of the Moody diagram in which the lines are horizontal. In rough pipes, therefore, head loss varies with the square of velocity.

Values for pipe roughness ϵ, taken from a paper by Lamont (1981), are presented in Table 2.1.

2.13. FRICTION FACTOR FOR TRANSITION FLOW

The smooth and rough flow equations are fairly easy to work with. However, the flow in most pipes in water distribution systems actually falls into a transition zone where the friction factor depends on both the relative roughness and Reynolds number. Colebrook and White developed the following formula for determining the friction factor in the transition zone:

$$f = \left[1.14 - 2 \log \left(\frac{\epsilon}{D} + \frac{9.35}{N_r \sqrt{f}} \right) \right]^{-2}. \qquad (2.13.1)$$

This expression approaches Eq. (2.11.2) as the pipe roughness becomes small, and approaches Eq. (2.12.1) as the roughness becomes large.

Another way of interpreting the Colebrook–White equation is in terms of

Table 2.1.   Experimental Data on Pipe Roughness.

Type of Pipe	Number of Experiments	Mean Value mm	Mean Value in.	Recommended Design Value mm	Recommended Design Value in.
Uncoated cast iron	3	0.226	0.0089	0.25	0.010
Coated cast iron	14	0.102	0.0040	0.125	0.005
Coated spun iron	5	0.056	0.0022	0.05	0.002
Galvanized iron	9	0.102	0.0040	0.125	0.005
Wrought iron	18	0.050	0.0020	0.05	0.002
Uncoated steel	11	0.028	0.0011	0.04	0.0015
Coated steel	6	0.056	0.0022	0.05	0.002
Uncoated asbestos-cement	13	0.028	0.0011	0.04	0.0015
Coated asbestos-cement	5	virtually smooth		smooth pipe	
Spun cement-lined—grade 1	6	smooth pipe		smooth pipe	
Spun cement-lined—grade 2	15	0.380	0.015	0.40	0.015
Spun bitumen-lined—grade 1	7	smooth pipe		smooth pipe	
Spun bitumen-lined—grade 2	5	0.120	0.0047	0.125	0.005
Smooth pipe*	18	smooth pipe		smooth pipe	
PVC pipe (with waviness)[†]	4	0.030	0.0012	0.04	0.0015
Prestressed concrete pipe (Freysinnet)	1	0.030	0.0012	0.04	0.0015
Spun concrete pipes (Bonna and Socoman)	10	0.200	0.0078	0.25	0.010
Pipes relined with cement mortar (Tate process)	17	0.510	0.020	0.500	0.020
Concrete miscellaneous Scobey[‡]	39				
Class 1—$C_s = 0.27$		5.10	0.200	5.00	0.200
Class 2—$C_s = 0.31$		1.27	0.050	1.25	0.050
Class 3—$C_s = 0.345$		0.41	0.016	0.50	0.020
Class 4—$C_s = 0.37$		0.18	0.007	0.25	0.010
Best—$C_s = 0.40$		0.102	0.004	0.125	0.005
Concrete tunnel linings Colebrook	18				
Best recorded		0.025	0.001		
Practical minimum		0.062	0.0025	0.062 to	0.0025 to
Mean		0.310	0.012	0.310	0.012
Probable maximum		1.55	0.060		

*Smooth pipes include smooth-drawn nonferrous pipes of aluminum, brass, copper, lead, and nonmetallic pipes of glass, polythene, and best quality PVC (free from waviness).
†PVC pipes with waviness (11.4 cm [4.5 in.] to 30.5 cm [12 in.]) as tested by G. Tison of the University of Ghent.
‡Scobey's class 1—old concrete pipes, mortar not wiped from joints; class 2—modern dry mix pipes, monolithic pipes, or tunnel linings made over rough forms; class 3—small wet mix pipes in short lengths, dry mix pipes in long lengths, average monolithic pipes made on steel forms, pressure-made pipes with interior coat of neat cement by mechanical trowel; class 4—monolithic pipes with joint scars and all irregularities removed. First class concrete against oiled steel forms. Best (added subsequently by Scobey)—premium pipes for municipal pipelines conveying reasonably clear water. Smooth concrete pipes produced by vibration in rigid oiled steel forms or by centrifugal spinning under definite specifications.

Exerpted from *Journal AWWA*, Vol. 73, No. 5 (May 1981), by permission. Copyright © 1981, The American Water Works Association.

boundary layer thickness. Eq. (2.13.1) can be rearranged as

$$f = \left\{ 1.14 - 2 \log \left[\frac{\epsilon}{D} \left(1 + \frac{9.35D}{\epsilon N_r \sqrt{f}} \right) \right] \right\}^{-2}. \qquad (2.13.2)$$

The $9.35D/\epsilon N_r \sqrt{f}$ term determines if the flow is smooth or rough. If this term is much less than one the flow is rough, while if it is much greater than one the flow is smooth. Substituting $N_r = VD/v$ and $\sqrt{f} = \sqrt{8} v^*/V$ and noting that the thickness of the laminar boundary layer can be approximated by $\delta = 10v/v^*$ gives

$$f = \left\{ 1.14 - 2 \log \left[\frac{\epsilon}{D} \left(1 + 0.33 \frac{\delta}{\epsilon} \right) \right] \right\}^{-2} \qquad (2.13.3)$$

where δ = thickness of laminar boundary layer, L. Eq. (2.13.3) shows that it is actually the relative height of the roughness elements as compared to the thickness of the laminar boundary layer that determines whether the flow is smooth or rough.

2.14. EXPLICIT FORMULAS FOR DETERMINING FRICTION FACTOR

While the Colebrook–White formula can be used to calculate the friction factor in turbulent flow, it is difficult to use since f must be calculated by trial and error. That is, the engineer must choose an f value, calculate a new f, and repeat the process until the procedure converges. Numerical methods such as the Newton–Raphson method can be used to speed convergence but the process is still somewhat time consuming unless it is incorporated into a computerized procedure. On the other hand, the Moody diagram is easy to use for manual calculations, but cannot be incorporated into a computerized procedure. Several approximations to the Colebrook–White equation have been developed to enable the engineer to solve explicitly for f. Wood (1966) uses a combination of power functions to determine f as

$$f = a + bN_r^{-c} \qquad (2.14.1)$$

where

$$a = 0.094(\epsilon/D)^{0.225} + 0.53(\epsilon/D)$$
$$b = 88(\epsilon/D)^{0.44}$$
$$c = 1.62(\epsilon/D)^{0.134}$$

For large Reynolds numbers the a term predominates, while for small Reynolds numbers it is negligible.

Swamee and Jain (1976) produced the following equation, which looks more like the Colebrook–White equation:

$$f = \frac{0.25}{\left[\log\left(\dfrac{0.27\epsilon}{D} + \dfrac{5.74}{N_r^{0.9}}\right)\right]^2}. \qquad (2.14.2)$$

The engineer must be careful in solving the Swamee and Jain formula for ϵ given f and N_r. In taking the square root of the denominator in Eq. (2.14.2), the negative root must be used.

Olujic (1981) gives some other explicit formulas used by chemical engineers. In general these formulas are sufficiently accurate for most problems, although some may break down for extreme values of f and N_r.

EXAMPLE. Calculate the friction factor for a pipe with relative roughness 0.0001 and a Reynolds number of 10^6 using (1) the Moody diagram, (2) the Colebrook–White formula, (3) the Wood formula, and (4) the Swamee and Jain formula.

The Moody diagram gives 0.0135.

The Colebrook–White formula gives

$$f = [1.14 - 2 \log (10^{-4} + 9 \times 10^{-6}/\sqrt{f})]^{-2}.$$

A typical initial guess for f is 0.02, which gives an f of 0.0132. Substituting back into the equation for f and repeating the calculation gives 0.0134 and 0.0134, which is the solution for f.

The Wood formula gives

$$a = 0.0119$$
$$b = 1.53$$
$$c = 0.47$$

and hence

$$f = 0.0119 + 1.53(10^6)^{-0.47} = 0.0142.$$

The Swamee and Jain formula gives

$$f = \frac{0.25}{\left[\log\left(0.27 \times 10^{-4} + \dfrac{5.74}{(10^6)^{0.9}}\right)\right]^2} = 0.0135.$$

2.15. EMPIRICAL HEAD LOSS EQUATIONS

One of the primary problems with the Darcy–Weisbach equation is that the friction factor is not a constant for a given pipe (except for pipes for which the flow is always rough). Several formulas have been presented, however, which contains constants that are alleged to be independent of the Reynolds number. Actually, the constants in these equations are only constant in specific instances, so the engineer must be cautious in applying such expressions.

In these equations, the friction factor is approximated by a formula of the form

$$f = KV^a / D^b \qquad (2.15.1)$$

where a, b, K are constants. The parameter K contains some other parameter which is an indicator of the roughness (or smoothness) of the pipe. Substitution of f into the Darcy–Weisbach equation gives

$$S = \left(\frac{K}{D^{1+b}} \right) \frac{V^{2+a}}{2g}. \qquad (2.15.2)$$

Some of the more commonly used equations are summarized in Table 2.2. The equations for which $a = 0$ are rough pipe equations which approach the shape of the curves in the Moody diagram only in the rough pipe range. Examples are the Manning equation and the Scobey equation for concrete pipe. On the other hand, the Hazen–Williams equation and the Scobey equation for steel pipe correspond to friction factors lying parallel to the smooth flow line of the Moody diagram.

For water distribution system analyses the most commonly used of the empirical formulas is the Hazen–Williams equation. Most water system engineers have a very good feel for the meaning of the Hazen–Williams C

Table 2.2. Summary of Empirical Head Loss Equations.

	Equation for S	Equation for V	K (in English units)*	a	b
Hazen–Williams	$\dfrac{3.03}{D^{1.16}} \left(\dfrac{V}{C} \right)^{1.85}$	$0.55CD^{0.63} S^{0.54}$	$\dfrac{194.5}{C^{1.85}}$	-0.15	0.16
Manning	$\dfrac{2.86n^2}{D^{1.33}} V^2$	$\dfrac{0.589}{n} D^{0.667} S^{0.50}$	$184.1n^2$	0	0.33
Scobey					
concrete	$\dfrac{4.46 \times 10^{-5} \, V^2}{C_s^2 \, D^{1.25}}$	$149.7 C_s D^{0.625} S^{0.50}$	$\dfrac{0.0029}{C_s^2}$	0	0.25
steel	$\dfrac{0.0010 K_s V^{1.9}}{D^{1.1}}$	$\dfrac{38.0 D^{0.58} S^{0.526}}{K_s^{0.526}}$	$0.0644 K_s$	-0.10	0.10

*Velocity in feet/second and diameter in feet.

factor, while pipe roughness remains a mystery to many practicing engineers. For this reason, the application of the Hazen–Williams equation is explained in more detail in the following section.

2.16. HAZEN–WILLIAMS EQUATION

For water distribution system applications, the most popular of the head loss equations is the Hazen–Williams equation. It is most commonly written

$$V = 0.55CD^{0.63}S^{0.54} \qquad (2.16.1)$$

where C = Hazen–Williams C factor. (In the above equation D is in feet and V is in feet/second. If V is in meters/second and D is in meters, then the 0.55 becomes 0.355.) Virtually any practicing engineer in the water field knows that a C of 140 is indicative of very smooth new pipe; old pipe in good condition has a C in the range 100 to 120; and pipe with tuberculation or heavy scaling have C factors of 80 to as low as 40. Table 2.3 taken from Lamont (1981) gives numerical values of C factors for a wide array of situations.

Table 2.3. Values of C in Hazen–Williams Formula.

Type of Pipe	2.5 cm (1 in.)	7.6 cm (3 in.)	15.2 cm (6 in.)	30.5 cm (12 in.)	61 cm (24 in.)	122 cm (48 in.)
			C Values for Certain Pipe Diameters			
Uncoated cast iron: smooth and new		121	125	130	132	134
Coated cast iron: smooth and new		129	133	138	140	141
30 years old						
Trend 1—slight attack		100	106	112	117	120
Trend 2—moderate attack		83	90	97	102	107
Trend 3—appreciable attack		59	70	78	83	89
Trend 4—severe attack		41	50	58	66	73
60 years old						
Trend 1—slight attack		90	97	102	107	112
Trend 2—moderate attack		69	79	85	92	96
Trend 3—appreciable attack		49	58	66	72	78
Trend 4—severe attack		30	39	48	56	62
100 years old						
Trend 1—slight attack		81	89	95	100	104
Trend 2—moderate attack		61	70	78	83	89
Trend 3—appreciable attack		40	49	57	64	71
Trend 4—severe attack		21	30	39	46	51
Miscellaneous						
Newly scraped mains		109	116	121	125	127
Newly brushed mains		97	104	108	112	115

Table 2.3. *(Continued)*

Type of Pipe	C Values for Certain Pipe Diameters					
	2.5 cm (1 in.)	7.6 cm (3 in.)	15.2 cm (6 in.)	30.5 cm (12 in.)	61 cm (24 in.)	122 cm (48 in.)
Coated spun iron—smooth and new		137	142	145	148	148
Old—take as coated cast iron of same age						
Galvanized iron—smooth and new	120	129	133			
Wrought iron—smooth and new	129	137	142			
Coated steel—smooth and new	129	137	142	145	148	148
Uncoated steel—smooth and new	134	142	145	147	150	150
Coated asbestos-cement—clean		147	149	150	152	
Uncoated asbestos-cement—clean		142	145	147	150	
Spun cement-lined and spun bitumen-lined—clean		147	149	150	152	153
Smooth pipe (including lead, brass, copper, polythene, and smooth PVC)—clean	140	147	149	150	152	153
PVC wavy—clean	134	142	145	147	150	150
Concrete—Scobey						
Class 1—C_s = 0.27; clean		69	79	84	90	95
Class 2—C_s = 0.31; clean		95	102	106	110	113
Class 3—C_s = 0.345; clean		109	116	121	125	127
Class 4—C_s = 0.37; clean		121	125	130	132	134
Best—C_s = 0.40; clean		129	133	138	140	141
Tate relined pipes—clean		109	116	121	125	127
Prestressed concrete pipes—clean				147	150	150

*The above table has been compiled from an examination of 372 records. It is emphasized that the Hazen–Williams formula is not suitable for the coefficient C values appreciably below 100, but the values in the above table are approximately correct at a velocity of 0.9 m/s (3 ft/s).

For other velocities the following approximate corrections should be applied to the values of C in the table above.

Values of C at 0.9 m/s	Velocities Below 0.9 m/s for Each Halving, Re-halving of Velocity Relative to 0.9 m/s	Velocities Above 0.9 m/s for Each Doubling, Re-doubling of Velocity Relative to 0.9 m/s
C below 100	add 5 percent to C	subtract 5 percent from C
C from 100 to 130	add 3 percent to C	subtract 3 percent from C
C from 130 to 140	add 1 percent to C	subtract 1 percent from C
C above 140	subtract 1 percent from C	add 1 percent to C

Exerpted from *Journal AWWA*, Vol. 73, No. 5 (May 1981), by permission. Copyright © 1981, The American Water Works Association.

Figure 2.7 gives head loss versus Q and V for a wide range of flows and diameters. To use the figure, one must know two of the four parameters (Q, V, D, h/L). The correction factors for C other than 130 must be applied to h/L.

The Hazen–Williams equation C factor differs from other coefficients such as the friction factor and Manning's n in that it indicates the carrying capacity (smoothness) of the pipe. The head loss predicted by the Darcy–Weisbach and Hazen–Williams equations can be set equal to develop the following relationship between the C factor and the friction factor:

$$C = \frac{17.25}{f^{0.54}(VD)^{0.081}}. \tag{2.16.2}$$

Eq. (2.16.2) shows that C varies inversely as $f^{0.54}$ but that the relationship depends on the Reynolds number (actually on VD, since the Hazen–Williams equation does not account for changes in kinematic viscosity of the fluid, as it is appropriate only for water at temperature at which the C-factor tests were conducted).

Relating C to the friction factor is of little practical value, since the friction factor is more difficult to determine than C. The friction factor is a function of the relative roughness of the pipe, which is a constant for a given pipe. Substituting for f in Eq. (2.16.2) using the Swamee and Jain formula gives

$$C = \frac{17.25\left[1.14 - 2\log\left(\dfrac{\epsilon}{D} + \dfrac{6.74 \times 10^{-4}}{(VD)^{0.9}}\right)\right]^{1.08}}{(VD)^{0.081}} \tag{2.16.3}$$

where

ϵ/D = relative roughness
V = velocity at which C factor is required, ft/sec
D = pipe diameter, ft.

The relative roughness can be determined indirectly by measuring the head loss and velocity in the pipe, calculating f or C and finding the roughness by solving one of the friction factor formulas for roughness, or by using

$$\frac{\epsilon}{D} = -\frac{6.72 \times 10^{-4}}{(V_0D)^{0.9}} + \text{antilog} \left((0.5)\{1.14 - [0.058C_0(V_0D)^{0.08}]^{0.926}\}\right) \tag{2.16.4}$$

where

V_0 = velocity at which tests were conducted, ft/sec
C_0 = C factor corresponding to velocity V_0.

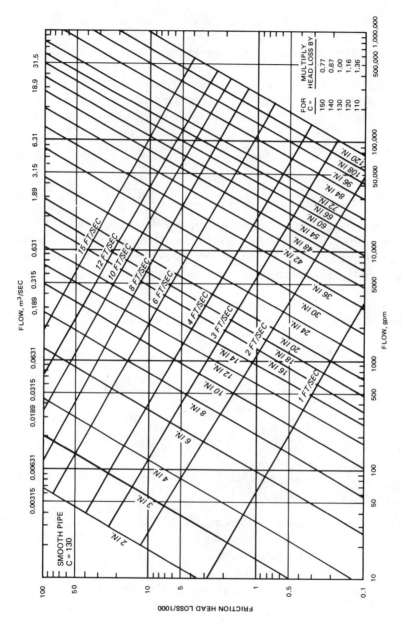

Fig. 2.7. Head loss in smooth pipe.

Eq. (2.16.4) was used to generate Fig. 2.8, which shows the relationship of roughness to C factor for several values of D for $V_0 = 3$ ft/sec.

Knowing the C factor and the velocity at which it was calculated, it is possible to determine the pipe roughness and a C factor corresponding to some other velocity. The important question is: how much difference does failing to correct the C factor for velocity make? Eqs. (2.16.3) and (2.16.4) appear imposing, and if the difference in C is small, it may not be worthwhile to make the correction. The footnote to Table 2.3 gives an indication of the magnitude of this effect. Doubling the velocity of a rough pipe decreases the C factor by 5 percent and increases the head loss by 10 percent (since head loss varies with C to the 1.85 power). For smooth pipes the effect is much less dramatic because, as stated previously, the Hazen–Williams equation is a much better model of smooth pipe flow than of rough pipe flow.

Therefore, there is no problem in using the Hazen–Williams equation for smooth flow or for other flows for which the velocity under consideration is close to the velocity at which C was measured. In other cases, C should be corrected by calculating the roughness and applying Eq. (2.16.3), especially if the velocity is expected to differ by more than a factor of 2 from the velocity at which C was measured.

The next question raised by most engineers is: how rough is a rough pipe in terms of a C factor? Fig. 2.9 shows that for the flow velocities and pipe diameters normally found in water distribution systems, smooth flow exists only at C values greater than 150, a rarity in distribution systems.

In the case of wholly rough pipes, Eq. (2.16.4) can be substituted into (2.16.3) and the terms including the Reynolds number can be eliminated to give

$$\frac{C}{C_0} = \left(\frac{V_0}{V}\right)^{0.081} \tag{2.16.5}$$

Using Eq. (2.16.5) it is relatively easy to correct the C factor for any velocity other than the one at which it was originally measured. Substituting Eq. (2.16.5) into Eq. (2.16.1) gives a rough flow Hazen–Williams equation of

$$V = 0.57 C_0^{0.925} S^{0.5} D^{0.58} V_0^{0.075}. \tag{2.16.6}$$

Solving for S gives

$$S = \frac{3.03 V^2}{C_0^{1.85} D^{1.165} V_0^{0.15}}. \tag{2.16.7}$$

This formula is actually more appropriate for use in water distribution

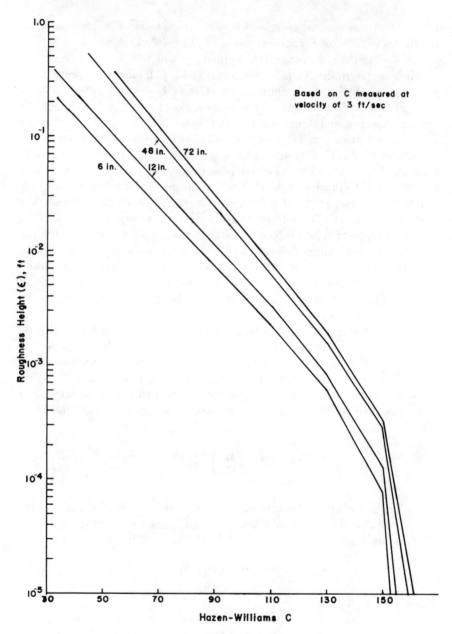

Fig. 2.8. Relationship between pipe roughness and *C* factor.

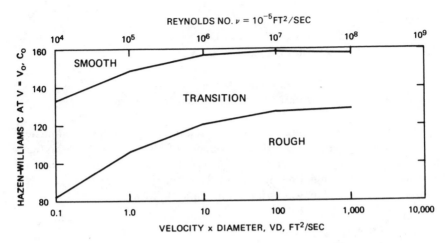

Fig. 2.9. C factors corresponding to smooth and rough flow.

system problems than the traditional Hazen–Williams equation, since in many systems, especially older systems, the flow is more likely to be rough than smooth. The formula requires knowing V_0, which is usually not reported with literature values of C. The problem is further complicated when actual and nominal pipe diameters are mixed. The problems involved with measuring and reporting the results of head loss tests are described in Chapter 8.

For many practical problems involving water distribution systems, the pipe diameter is known in inches, the flow in gallons per minute, and the head loss is desired in feet per 1000 feet of length. The Hazen–Williams equation can be given for use with these units as follows:

Smooth:

$$S_{1000} = \frac{10{,}460}{D_i^{4.87}} \left(\frac{Q}{C}\right)^{1.85} \qquad (2.16.8)$$

Rough:

$$S_{1000} = \frac{10{,}460 Q^2}{D_i^{4.87} Q_0^{0.15} C_0^{1.85}} \qquad (2.16.9)$$

or, for rough flow when C was measured at a velocity of 3 ft/sec,

$$S_{1000} = \frac{7780 Q^2}{D_i^{5.16} C_0^{1.85}} \qquad (2.16.10)$$

where

$$S_{1000} = \text{head loss, } L/1000L$$
$$Q = \text{flow, gpm}$$
$$D_i = \text{diameter, in.}$$
$$C_0 = \text{measured } C \text{ factor}$$
$$Q_0 = \text{flow at which } C \text{ was measured, gpm.}$$

For quick reference, Fig. 2.10 gives the relationship between flow and head loss for rough pipes with $C = 100$ at a velocity of 3 ft/sec. Note that this figure is not simply Fig. 2.7 for a lower C factor. Instead the head loss lines are steeper, since in rough flow head loss varies with flow (and velocity) to the 2 rather than the 1.85 power as it does for smooth flow.

2.17.   MINOR LOSSES

The expression "minor losses" is used to describe energy losses caused by valves, bends and changes in pipe diameter. These energy losses are generally smaller than friction losses in straight sections of pipe in water distribution systems and are usually ignored in practical problem solving. There are instances, however, where these minor losses become significant and must be considered. Minor losses occur because valves and fittings create turbulence in excess of that produced in a straight pipe. Since this energy cannot be recovered by the fluid, it is lost.

Minor losses can be expressed in three ways:

1. A minor loss coefficient K may be used to give head loss as a function of velocity head

$$h = K \frac{V^2}{2g} \tag{2.17.1}$$

where K = minor loss coefficient.

2. Minor losses may be expressed in terms of the equivalent length of straight pipe, or as pipe diameters (L/D) which produces the same head loss. For example, head loss may be calculated by the Darcy–Weisbach equation

$$h = f \frac{LV^2}{D2g} \tag{2.17.2}$$

where f = friction factor of straight pipe. The equivalent pipe length and

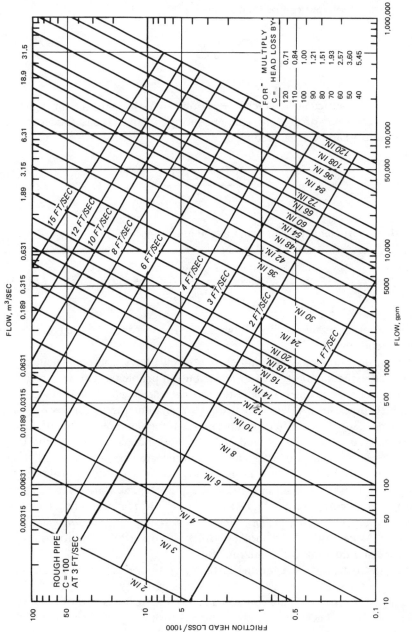

Fig. 2.10. Head loss in rough pipe.

minor loss coefficient can be related by

$$\frac{L}{D} = \frac{K}{f}. \tag{2.17.3}$$

3. A flow coefficient C_v which gives a flow (gpm) that will pass through the valve at a pressure drop of 1 psi may be specified. Given the flow coefficient the head loss can be calculated as

$$h = \frac{18.5 \times 10^6 D^4 V^2}{C_v^2 2g}. \tag{2.17.4}$$

The flow coefficient can be related to the minor loss coefficient by

$$K = \frac{18.5 \times 10^6 D^2}{C_v^2} \tag{2.17.5}$$

where D = diameter, ft. The flow coefficient is generally used by value manufacturers.

Of the formulations for minor losses given above, the equivalent pipe approach is the handiest for water distribution problems. This is especially true for network problems where the engineer need only add a few feet of length to pipes near valves and fittings to account for minor losses. The approach using flow coefficients is the most cumbersome, as flow coefficients have units of gpm and depend on the pipe diameter. The minor loss coefficients and equivalent lengths are dimensionless and only slightly dependent on diameter. Values for minor loss coefficients and equivalent pipe lengths for most typical water distribution appurtenances are given in Table 2.4. For more comprehensive tables and figures, the reader is referred to *The Handbook of Valves, Piping and Pipelines* (Warring, 1982) or *Internal Flow Systems* (Miller, 1973). The book by Miller also describes such situations as diffusers and dividing and combining flows.

Not all of the head loss occurs at the valve or fitting, as the turbulence created takes some distance before it is dissipated by viscous forces. For example, if several bends are connected in series to form a coil, the head losses cannot be taken because the sum of the losses of each bend as some of the turbulence caused by each bend tends to be masked by the turbulence of the preceeding and succeeding bends.

Table 2.4.   Minor Losses.

	Loss Coefficient (K)	Equivalent Length (L/D)
Gate valve—open	0.39	13
¾ open	1.10	35
½ open	4.8	160
¼ open	27.	900
Globe valve—open	10.	350
Angle valve—open	4.3	170
Check valve—conventional	4.0	130
Check valve—clearway	1.5	50
Check valve—ball	4.5	150
Butterfly valve—open	1.2	40
Cock—straight through	0.5	18
Foot valve—hinged	2.2	75
Foot valve—poppet	12.5	420
90 deg std. elbow	0.9	30
45 deg std. elbow	0.45	15
Standard tee—flow through run	0.6	20
Standard tee—flow through branch	1.8	60
Return bend	1.5	50
90 deg bend		
$r/d = 2$	0.3	9
$r/d = 8$	0.4	12
$r/d = 20$	0.5	18
(r = bend radius, d = pipe diameter)		
Mitre bend		
90 deg	1.8	60
60 deg	0.75	25
30 deg	0.25	8
(Mitre losses include pipe one diameter up and downstream)		
Expansion		
$d/D = 0.75$	0.18	6
$d/D = 0.5$	0.55	18
$d/D = 0.25$	0.88	29
Contraction		
$d/D = 0.75$	0.18	6
$d/D = 0.5$	0.33	10
$d/D = 0.25$	0.43	14
(Velocities for use with sudden contraction and expansion refer to smaller pipe)		
Entrance—projecting	0.78	26
Entrance—sharp	0.50	17
Entrance—well rounded	0.04	1
Exit	1.0	33

REVIEW QUESTIONS

1. Why can the kinetic energy correction factor be ignored in most water distribution problems?
2. When can the Hazen–Williams equation be applied to problems involving rough flow?
3. Why is the expression "rough flow" more appropriate than "rough pipe" in Question 2?
4. How does temperature affect head loss in a water distribution system?
5. For a valve, which varies more with size, K or C_v?
6. The Reynolds number gives an indication of the relative size of _____ forces to inertial forces.
7. What is eddy viscosity a measure of?
8. For a given flow rate, head loss (in rough flow) is proportional to pipe diameter to what power? (i.e., if $S = \text{const}/D^x$, what is x?
9. Is (v_{max}/V) larger in laminar or turbulent flow? Why?
10. Are smooth pipes in turbulent flow represented by a single line or many lines on a Moody diagram? Why?
11. What is the advantage of the Swamee and Jain formula over the Colebrook–White formula?
12. What are the advantages and disadvantages of the Hazen–Williams equation when compared with the Darcy–Weisbach equation?
13. What are typical values of the Hazen–Williams C factor and pipe roughness for:
    a. Smooth pipe?
    b. New pipe?
    c. Old rough pipe?
14. Is eddy viscosity more important in laminar or turbulent flow?
15. Does neglecting minor losses cause less error in long pipelines or complex manifolds?
16. The Blasius equation can be used for (turbulent/laminar/both) flow for (smooth/rough/transition) flow at (high/low/all) Reynolds numbers.
17. In a Newtonian fluid, viscous momentum transport is proportional to velocity gradient. What is the proportionality constant?
18. What happens to the boundary layer in rough flow?
19. In addition to the height of a roughness element, what other roughness characteristics of a pipe are important?
20. Is flow through a pipe having a Hazen–Williams C of 100 smooth, transition, or rough?

PROBLEMS

1. Prepare a plot of Hazen–Williams C factor versus log relative roughness for a Reynolds number of 10^7.
2. Given a pipe in which $Q = 1000$ gpm, $S = 20$ ft/1000 ft and nominal diameter is 12 in., find the friction factor (based on rough pipe), Hazen–Williams C, relative roughness, and roughness. Repeat the above exercise if the actual diameter is 10 in. (Use a kinematic viscosity of 10^{-5} ft^2/sec.)

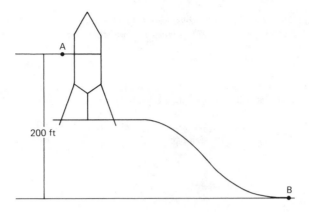

Fig. 2.1p. Problem 2.6.

3. A pump at point A produces 200 ft of head at a flow of 1000 gpm. If the head loss between points A and B is 20 feet, what is the pressure at B in psi and kPa if B is at the same elevation as the water surface of the reservoir from which the pump is withdrawing the water?

4. Find the kinetic energy correction factor for a 24-in.-diameter pipe with a centerline velocity of 2.4 ft/sec. Use the $\frac{1}{7}$th power law for the velocity profile.

5. Using the Hazen–Williams equation for smooth and rough flow, find the head loss that will occur in 350 ft of 18-in.-diameter pipe carrying 20 cfs with a C of 110 measured at 3 ft/sec.

6. What is the elevation of the hydraulic grade line and energy grade line at point B in Fig. 2.1p for $Q = 16.2$ cfs, given $D = 24$ in., $L = 2000$ ft, and $C = 130$? What are the elevations at zero flow?

7. For a smooth pipe, use the Blasius equation to show that the Hazen–Williams equation can be applied to a fluid other than water by correcting the C factor as follows:

$$C_o = C_w \left(\frac{\nu_o}{\nu_m} \right)^{0.081}$$

where

$$
\begin{aligned}
C_o &= C \text{ for other fluid} \\
C_w &= C \text{ for water} \\
\nu_o &= \text{kinematic viscosity of other fluid} \\
\nu_w &= \text{kinematic viscosity of water.}
\end{aligned}
$$

REFERENCES

Bird, R.B., W.E. Stewart and E.N. Lightfoot, 1960, *Transport Phenomena,* John Wiley & Sons, Inc., New York.

Lamont, P.A., 1981, Common Pipe Flow Formulas Compared with the Theory of Roughness, *J. AWWA,* Vol. 73, No. 5, p. 274.

Miller, D.S., 1978, *Internal Flow Systems,* BHRA Fluid Engineering, Great Britan.

Morris, H.M. and J.M. Wiggert, 1972. *Applied Hydraulics in Engineering,* The Ronald Press, New York.

Olujic, Z., 1981, Compute Friction Factors Fast for Pipe Flow, *Chemical Engineering,* Dec. 14, p. 91.

Schlichting, H., 1979, *Boundary-Layer Theory,* McGraw-Hill, New York.

Swamee, P.K., and A.K. Jain, 1976, Explicit Equation for Pipe Flow Problems, *J. ASCE Hyd. Div.,* Vol. 102, No. HY5, p. 657.

Waring, R.H., 1982, *Handbook, of Valves, Piping and Pipelines,* Gulf Publishing Co.

Wood, D.J., 1966, An Explicit Friction Factor Relationship, *Civil Engineering,* Vol. 32, No. 12, p. 60.

3 | SOLVING PIPE NETWORK FLOW PROBLEMS

3.1. INTRODUCTION

The previous chapter described the basics of closed conduit hydraulics. Real water distribution systems, however, do not consist of a single pipe, but complicated combinations of pipes, pumps and other appurtenances, which are referred to as a *network*. Solving water network problems involves judicious application of the continuity and energy equations presented in Chapter 2 to simplify them to a level of complexity that can be solved by computer and in some cases manually.

The type of solution technique required depends to a great extent on whether the network is branched or looped. Looped systems require an extra equation to enable problems to be solved (i.e., net head loss around a loop is zero). Systems with pumps require an extra equation for each pump.

Before specific techniques for solution of network problems are described, some rules for using equivalent pipe techniques to simplify problems are presented. Methods for handling pumps and valves are also discussed. Once the background material has been presented, methods for actually solving network problems are given. Generally, the solution techniques consist of setting up a continuity equation for each node (a node is the intersection of two or more pipes), an energy equation for each pipe, and special equations for tanks, pumps, and valves, and then combining these equations to make their solution practicable.

In general, pipe network problems can be categorized by: (1) the manner in which time is considered in the analysis (steady, gradually varied, transient); (2) the topology of the system (series, branched, branched with multiple tanks, looped); (3) the way in which the equations are set up (flow equations, head equations, loop equations, optimization); and (4) the method used to solve the resulting set of equations (analytical, graphical, Hardy–Cross, linear theory, Newton–Raphson).

3.2. TYPES OF PROBLEM

In the head loss equations in the previous chapter, five variables were included in each equation: length, diameter, flow, head loss, and roughness (or carrying capacity). In general, the pipe length and roughness are fixed beforehand, so the engineer is usually solving for one of the other three variables. Finding flow and head (or head loss) is basically a hydraulic problem. On the other hand, finding the best pipe diameter calls for both hydraulic and economic consideration. Problems of this type will be given detailed attention in Chapter 5.

In most network problems, flow rates in individual pipes are generally not known at the outset; rather, only the outflows from the network and head at one or more points are given. A solution consists of determining the flow in each pipe and the head at each node in the system. To calculate the head at various points in the system it is necessary to know the head at one or more points beforehand; otherwise only relative heads can be determined.

A network problem is considered solved when the flow in each pipe and head at each node has been determined with the desired accuracy. (In most water distribution systems, the difference between total head and static or piezometric head is negligible.)

3.3. TYPES OF NETWORK

All pipe network problems can be divided into two important categories depending on whether flow in individual pipes can be determined without the need to solve the energy equation. In problems in which the head is fixed at only one location and there are no loops, it is possible, given the outflows and inflows to the system, to determine the flow in each pipe without solving the energy equation. For this kind of network, the continuity equation can be solved to yield flows in individual pipes, and then the energy equation can be used to calculate the heads. Such network problems can be solved manually, since the individual equations can be solved one at a time.

If there is a loop in the network, or if the head is specified at more than one location, then the energy and continuity equations cannot be solved independently and problems become much more difficult. Such problems are usually referred to as *looped network problems,* although some very simple networks not involving loops also fall into this category. For example, a single pipe with head known at both ends falls into this category, since it is necessary to solve the energy equation (and a trivial continuity equation) to know the flow in the pipe and the pressure along the pipe. Networks for which head is fixed at more than one point can be considered to have a "pseudo-loop" consisting of the pipes connecting the two constant head points, plus an imaginary pipe with head loss exactly equal to the difference in head between the two points. Except for very simple cases involving one or two pipes, these problems must be solved numerically, usually with a computer.

For the remainder of this text, the expression *looped network* will refer to networks with either actual or pseudo-loops, and the term *branched network* will refer to networks with no loops and one or no fixed head point.

3.4. ENERGY EQUATION FOR LOOPED SYSTEM

In Chapter 2, the energy equation was written between two points along a pipe. For a pipe loop, the beginning and ending points are the same, and since head cannot take on two different values at the same point, the head at the beginning and ending of the loop are the same. Therefore, Eq. (2.6.1) must be written as follows, since points 1 and 2 are the same:

$$E_2 - E_1 = W - H = 0 \qquad (3.4.1)$$

where

$$E_1 = \text{energy at point 1}$$
$$E_2 = \text{energy at point 2}$$
$$W = \text{work done on fluid}$$
$$H = \text{head loss.}$$

This means that the net work done on the fluid in a loop must equal the head loss in the loop. It also means that flow cannot occur in one direction around a loop without a pump. In reality loops generally consist of several pipes, with or without pumps. Thus, the energy equation, written in units of head, becomes

$$\Sigma h_{pi} - \Sigma h_i = 0 \qquad (3.4.2)$$

where

$$h_{pi} = \text{head provided by } i\text{th pump, } L$$
$$h_i = \text{head loss in } i\text{th pipe, } L.$$

To use Eq. (3.4.2), it is necessary to know (or assume) the direction of flow in each pipe. Thus, it is convenient to write the head loss equations such as Darcy–Weisbach or Hazen–Williams as

$$h = KV|V|^{n-1} \qquad (3.4.3)$$

where

$$K = \text{constant in head loss equation}$$
$$n = \text{exponent in head loss equation}$$
$$V = \text{velocity in pipe, } L/T.$$

(The K in Eq. (3.4.3) is not the same K as in equation (2.15.1).)

Using an equation of this form enables the engineer to keep track of the direction of flow. The use of the parameters K and n makes the head loss equation very general. For example, with $n = 1.85$, it is the smooth flow Hazen–Williams equation, while for $n = 2$, it is the Darcy–Weisbach or Hazen–Williams equation, modified for rough pipe. The K parameter simply lumps the remainder of the terms which are essentially independent of velocity. For the Darcy–Weisbach equation, $K = fL/2gD$. In many of the problems presented in this chapter, it is more convenient to write Equation (3.4.3) in terms of flow rate rather than velocity. This only changes the numerical value of K (since for this case, it must contain the pipe area) but does not change the applicability of the equation.

3.5. EQUIVALENT PIPES

In solving pipe network problems, especially manually, the engineer can often save time by replacing groups of pipes with a single equivalent pipe. In order to replace a group of pipes by a single equivalent pipe between two points in a network, say A and B, the flow past A must be the same as the flow past B and there must be no inflows or outflows between the points. In a real water distribution system, some use or loss will occur along almost all pipes. However, the equivalent pipe approach can still be used as long as the use or loss is small in comparison with the total flow.

The problem of determining the characteristics of the equivalent pipe is essentially one of calculating K_e and n such that $K_e Q^n$ will predict the same head loss (for all flows) that would be determined by calculating the head loss for each pipe and combining them in an appropriate way. The use of equivalent pipes eliminates the need to calculate how the total flow is distributed among the pipes under consideration.

For pipes connected in series, the head losses are additive. That is

$$K_e Q_e^n = K_1 Q_1^n + K_2 Q_2^n + \cdots + K_N Q_N^n. \tag{3.5.1}$$

Note that all of the n's are treated as being the same. While this may not actually be the case, the difference in head loss will be small and not worth the extra effort required to account for variation in the n values. Since all the Q's are the same (i.e., no significant inflows or outflows), it is possible to divide through by Q_e^n to give

$$K_e = K_1 + K_2 + \cdots + K_N. \qquad (3.5.2)$$

Therefore to determine the head loss coefficient K_e for pipes in series, it is only necessary to add the individual K values.

In entering data for most pipe network computer programs, the engineer is not requested to enter values for K, but for diameter, length, and carrying capacity (or roughness). The engineer must therefore determine L_e, D_e, and in this case C_e for the equivalent pipe. For this purpose, Eq. (3.5.2) can be rewritten

$$\frac{L_e}{D_e^m C_e^n} = \frac{L_1}{D_1^m C_1^n} + \frac{L_2}{D_2^m C_2^n} + \cdots + \frac{L_N}{D_N^m C_N^n}. \qquad (3.5.3)$$

The logical choice for the length of the equivalent pipe is the sum of the lengths of the other pipes. Since changes in pipe diameter are generally larger than changes in C along a series of pipes, it is best to select a C factor for the equivalent pipe and calculate the diameter of the equivalent pipe. Dividing through by L_e and solving for D_e gives

$$D_e = \left[\frac{C_e^n}{L_e} \sum_{i=1}^{N} \left(\frac{L_i}{D_i^m C_i^n} \right) \right]^{-1/m} \qquad (3.5.4)$$

Thus, the length of the equivalent pipe is the sum of the length of the individual pipes and the diameter is a weighted average given by Eq. (3.5.4).

EXAMPLE. Find the length and diameter of an equivalent pipe to replace 1000 ft of 12-in. pipe and 3000 ft of 16-in. pipe connected in series, each having $C = 120$.
   The length is given by

$$L = 1000 + 3000 = 4000 \text{ ft}$$

The diameter (for $m = 4.87$, Hazen–Williams) is given by

$$D_e = \left[\frac{1}{4000} \left(\frac{1000}{12^{4.87}} + \frac{3000}{16^{4.87}} \right) \right]^{-0.205}$$

$$= 14.2 \text{ in.}$$

Note that if the diameter was calculated as the weighted average based on relative length, the equivalent diameter would have been 15 in. The error in head loss calculations would have been $(15/14.2)^{4.87}$, which is 31 percent.

To determine the characteristics of a pipe that is equivalent to a group of parallel pipes, it is necessary to recognize that the head loss is the same in each pipe:

$$h_e = h_1 = h_2 = \cdots = h_N. \tag{3.5.5}$$

To solve this for K_e, note that the flows in each individual pipe can be summed to give the flow in the equivalent pipe:

$$Q_e = Q_1 + Q_2 + \cdots + Q_N. \tag{3.5.6}$$

Substituting $(h/K)^{1/n}$ for each Q and noting that all of the head losses are equal gives

$$K_e = \left[\sum_{i=1}^{N} \left(\frac{1}{K_i} \right)^{1/n} \right]^{-n}. \tag{3.5.7}$$

Equation (3.5.7) must now be converted into such a form that the engineer can adjust either length, diameter, or C factor. One philosophy used in working with parallel pipes is to use the length and diameter of the largest pipe and adjust the C factor upward to account for the carrying capacity of the parallel pipes. Replacing the K terms in Eq. (3.5.7) with the appropriate terms from the Hazen–Williams equation and solving for the C of the equivalent pipe gives

$$C_e = C_1 + \sum_{i=1}^{N} C_i \left(\frac{D_i}{D_1} \right)^{m/n} \left(\frac{L_1}{L_i} \right)^{1/n}. \tag{3.5.8}$$

Equation (3.5.8) can be interpreted as stating that C_e is C_1 plus a correction factor for each additional pipe in parallel; that is,

$$C_e = C_1 + \sum_{i=1}^{N} \Delta C_i \tag{3.5.9}$$

where each of the ΔC_i terms is one of the terms in the summation in Eq. (3.5.8). The procedure for determining these ΔC_i values can be simplified greatly using the nomogram shown in Fig. 3.1. Unfortunately, using the same length

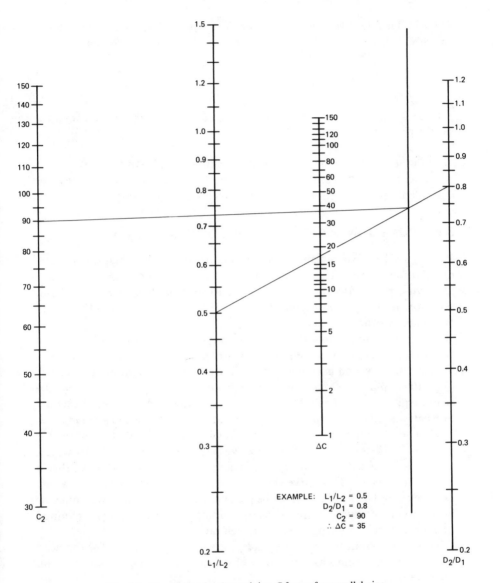

Fig. 3.1. Nomogram for determining C factor for parallel pipe.

and diameter of the largest pipe, and adjusting the friction factor does not work as well with the Darcy–Weisbach equation, since it may result in negative pipe roughnesses. For instance, a C of 200, which is possible when replacing a bundle of pipes with an equivalent pipe, would result in negative roughness for the equivalent pipe.

EXAMPLE. Find the C value for a pipe equivalent to parallel pipes with length 1000 ft and 2000 ft, diameters of 10 in. and 8 in., and C factors of 120 and 90 respectively.

$$\frac{L_e}{L_2} = 0.5, \quad \frac{D_e}{D_2} = 1.25$$

Figure 3.1 shows that the correction to the C of the larger pipe to account for the smaller pipe is 35, so that $C_e = 120 + 35 = 155$.

With the capability of modern day computers, simplifying a network with equivalent pipes is not as important as in the days of manual calculations, but equivalent pipes can still be helpful in reducing computer costs.

3.6. PUMPS IN WATER DISTRIBUTION SYSTEMS

Books have been written about pumps, pump stations, and their use in water distribution systems. This section deals with methods of representing pumps in solving pipe network flow problems. Readers interested in design and selection of pumps are referred to Sanks (1981), Karassik et al. (1976), and Hicks and Edwards (1971).

Virtually all pumps in water distribution systems are centrifugal pumps, so the discussion below is limited to that type of pump. Pump capacity is usually given as the flow and head delivered by the pump at its maximum efficiency (e.g., 1000 gpm at 200 ft). If the pump produces more or less flow, the efficiency will decrease. The relationship between flow and head is represented by what is called a *head characteristic curve*. A typical curve is shown in Fig. 3.2.

To determine the point on the curve at which the pump will operate, it is necessary to determine the resistance the pump must overcome. However, since this varies with the flow rate, it is necessary to develop what is called a *system head curve*. This curve shows the relationship between the flow rate and the combined head that must be produced to lift the water, overcome the head loss in the suction and discharge piping, and maintain the required pressures or raise water to the required elevation.

These curves can be derived by solving the energy equation for the head provided by the pump as a function of flow. Thus, systems head equation are developed from

$$h_p = z_2 - z_1 + \frac{P_2 - P_1}{\gamma} + h \qquad (3.6.1)$$

where

h_p = head required from pump, L

z_2 = elevation at downstream point, L

z_1 = elevation at upstream point, L
P_2 = pressure required at downstream point, M/LT^2
P_1 = pressure at upstream point, M/LT^2
γ = specific weight of fluid, M/L^3
h = head loss between points 1 and 2, L.

The pressure and elevation terms are independent of flow, but the head loss term h is highly dependent on flow. This gives the system head curve a positive slope. System head curves for some typical situations are shown in Fig. 3.3. Eq. (3.6.1) is often rewritten as follows:

$$h_p = h_L + h = h_L + KQ^n \tag{3.6.2}$$

where h_L = lift required, L. The last term in Eq. (3.6.2) corresponds to the last term in Eq. (3.6.1).

A pump or group of pumps will operate at a point at which the head produced is exactly offset by the lift and friction losses. This corresponds to a point that falls on both the head characteristic curve and the system head curve (i.e., the intersection of the two curves) as shown in Fig. 3.4.

It is also possible to determine the pump operating point analytically or numerically once a suitable equation can be found to represent the pump head

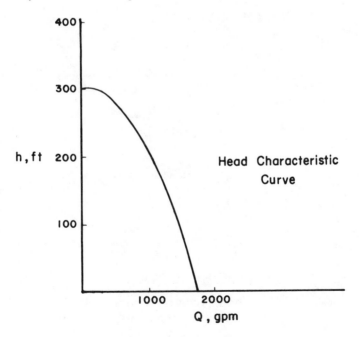

Fig. 3.2. Typical pump head characteristic curve.

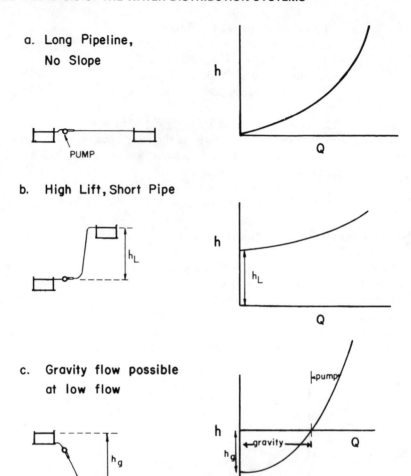

a. Long Pipeline, No Slope

PUMP

b. High Lift, Short Pipe

c. Gravity flow possible at low flow

Fig. 3.3. System head curves.

characteristic curve. The curve can best be represented by a parabola, as shown below:

$$h_p = h_0 + BQ + AQ^2 \qquad (3.6.3)$$

where

h_0 = head produced at no flow, L
A, B = regression coefficients.

The coefficients can be found by using a second degree polynomial regression

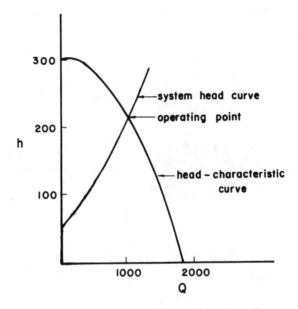

Fig. 3.4. Graphical determination of operating point.

method, but a simple approach can be developed knowing h_0 and two other points on the curve (Q_1, h_1) and (Q_2, h_2). B can be determined as

$$B = \frac{(h_2 - h_0) - (h_1 - h_0)(Q_2/Q_1)^2}{Q_2 - Q_1(Q_2/Q_1)^2}. \tag{3.6.4}$$

Once B is known, A can be determined as

$$A = (h_1 - h_0 - Q_1 B)/Q_1^2. \tag{3.6.5}$$

The coefficients can also be determined graphically by dividing Eq. (3.6.3) by Q and noting that a plot of Q versus $(h - h_0)/Q$ will yield a straight line with slope A and intercept B.

EXAMPLE. Given the head characteristic curve shown in Fig. 3.4, find the coefficients A, B, and h_0 to approximate the curve with a parabola. Use both the graphical and analytical method.

First note that $h_0 = 300$ ft and prepare a table of Q versus $(h - h_0)/Q$:

Q	h	$(h - h_0)/Q$
500	278.3	−0.043
1000	206.7	−0.093
1500	85.0	−0.143

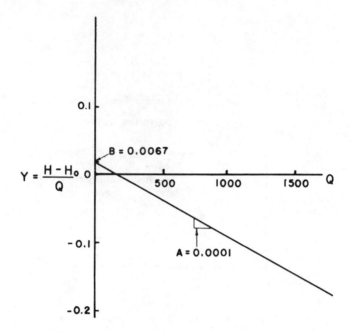

Fig. 3.5. Graphical method for determining pump head curve coefficients.

which is plotted in Fig. 3.5 to show $B = 0.0067$. The slope of the line can be calculated as

$$A = \frac{-0.040 - (-0.140)}{500 - 1500} = 0.00010.$$

Using Eq. (3.6.4) and (3.6.5) gives

$$B = \frac{(90 - 300) - (280 - 300)(3)^2}{1500 - 500(3)^2} = 0.0067$$

and

$$A = [210 - 300 - (1000)(0.0067)]/1000^2$$

$$= -0.00010.$$

Therefore

$$h_p = -0.0001Q^2 + 0.0067Q + 300.$$

To check, insert $Q = 500$ into the above equation to yield 278 ft, which is a reasonable value for head.

Once equations for the system head and pump characteristic curves are known, they can be set equal and solved to yield the flow and head at the operating point. This yields an equation of the form

$$h_0 + BQ + AQ^2 - h_L - KQ^n = F(Q) = 0. \qquad (3.6.6)$$

Solving this equation for Q can be a tedious process requiring a numerical solution. The Newton–Raphson method as applied to a single equation can be used to generate a new estimate of Q given a previous estimate as

$$Q_{i+1} = Q_i - F(Q_i)/F'(Q_i) \qquad (3.6.7)$$

where $Q_{i+1} = Q$ for the $(i + 1)$th iteration. Taking the derivative of Eq. (3.6.6) and substituting for F and F' in Eq. (3.6.7) gives

$$Q_{i+1} = Q - \frac{h_0 + BQ + AQ^2 - KQ^n - h_L}{B + 2AQ - KnQ^{n-1}}. \qquad (3.6.8)$$

While Eq. (3.6.8) looks quite messy, it converges to the correct value of Q quickly and can easily be used in programable calculator solutions. [Note that, for simplicity, subscript i has been omitted from the Q terms on the right side of Eq. (3.6.8).]

Once the flow has been determined using Eq. (3.6.8), the head can be calculated from the pump head characteristic curve or system head curve.

EXAMPLE. Given pump head curve coefficients of $h_0 = 300$, $B = 0.01$, and $A = -0.0001$ and system head curve coefficients of $K = 0.0004$, $n = 1.85$, and $h_L = 50$, find the flow and head produced by the pump.

Insert the appropriate values into Eq. (3.6.8) to give

$$Q_{i+1} = Q_i - \frac{300 + 0.01Q_i - 0.0001Q_i^2 - 0.0004Q_i^{1.85} - 50}{0.01 - 0.0002Q_i - 0.00074Q_i^{0.85}}.$$

Using $Q = 500$ as a starting point, the solution quickly converges as shown below:

Iteration	Q	F
1	500	191.
2	1309	-142.
3	1065	-12.4
4	1039	-0.140
5	1039	0.00002

Substituting 1039 gpm into the system head curve gives

$$h = 0.0004(1039)^{1.85} + 50 = 202 \text{ ft.}$$

To check this result, substitute for Q in the pump head equation:

$$h_p = (-0.0001)(1039)^2 + 0.01Q + 300 = 202.4 \text{ ft.}$$

In applying the Newton–Raphson method it is important to use a reasonable initial estimate for flow, otherwise the solution may converge slowly or not at all. For example, if 1000 gpm had been used as the starting point, the solution would have converged in 2 iterations, while if 0 gpm had been used, the solution would have never been reached because of a turning point in the pump characteristic curve near zero. Thus, if a solution should behave erratically after a few iterations, it is best to start over with a new initial estimate.

3.7. COMBINING PUMPS

Very seldom does a water distribution system rely on a single pump. Instead, pumping stations are usually equipped with several pumps operating in parallel. Selecting the ideal number of pumps for a given situation involves a tradeoff between cost and reliability.

For example, a single pump will pump a given flow at the least cost, but with no redundancy if it should fail to operate for some reason. Two pumps each having half the total required capacity cost more but provide greater reliability. Three pumps would provide more reliability but eventually the cost of adding another pump is not offset by the gain in reliability.

When two or more pumps are running in parallel, the flow that they produce is additive and the heads are equal (i.e., the same concept as in parallel pipes). Since the distance the flow must travel in the manifold connecting the downstream ends of the pumps is relatively small, the head produced by each pump must be identical, otherwise they would be pumping into one another.

In general, the easiest way to find the operating point for several pumps operating in parallel is to use a graphical solution. It is important to realize that to develop a head characteristic curve for pumps in parallel, the flows and not the heads must be added. Fig. 3.6 shows the head characteristic curve for pumps A and B added in parallel.

EXAMPLE. Find the operating point for pumps A and B operating individually and in parallel in a pumping station, given the head characteristic data in the table below. The

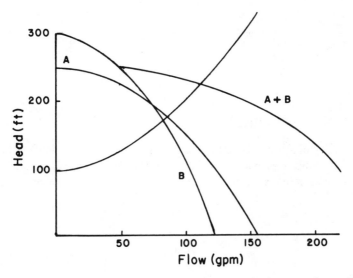

Fig. 3.6. Graphical method for determining operating point for pumps in parallel.

system head curve can be given by

$$h = 100 + 0.010 \, Q^2.$$

	h (ft)	
Q (gpm)	Pump A	Pump B
0	250	300
50	225	250
100	150	100
120	106	12

First plot the pump curves for each pump. Then add the flows graphically or prepare a table in which h is the independent variable as given below:

	Q (gpm)		
h (ft)	Pump A	Pump B	Pump A + B
100	122	100	222
150	100	87	187
200	70	71	141
250	0	50	50

Using the data in this table, it is possible to plot the A + B curve. Now plotting the system head curve and finding the operating points graphically yields:

	Q	h
A only	87	175
B only	82	166
A and B	110	225
	(A-45, B-65)	

3.8. PUMP EFFICIENCY

The efficiency of a pump is defined as the ratio of the power delivered by the pump to the power provided the pump by the driver, which in most water distribution systems is an electric motor. This ratio can be written mathematically as

$$e_p = \frac{\text{whp}}{\text{bhp}} \qquad (3.8.1)$$

where

e_p = efficiency
whp = water horsepower, ML^2/T^3
bhp = brake horsepower of driver, ML^2/T^3.

The water horsepower is the product of the flow rate and the head, converted into horsepower units as

$$\text{whp} = Qh/8.81 \qquad (3.8.2)$$

where

whp = water horsepower, hp
Q = flow, cfs
h = head, ft.

When the flow is given in gallons per minute, the constant in Eq. (3.8.2) is 3960, instead of 8.81.

The efficiency is usually multiplied by 100% and expressed as a percentage. Typical pumps in a water distribution system operate with a pump efficiency on the order of 70%. When the a pump is described in terms of its capacity and head (e.g., 300 ft at 200 gpm), the capacity and head are those at the point on the head characteristic curve with the maximum efficiency.

In working with pumps there are two other efficiencies that must be considered. The first is the driver efficiency which measures how effectively the driver converts electrical power (usually referred to as "wire horsepower") into mechanical energy. It is defined as

$$e_d = \frac{bhp}{khp} \qquad (3.8.3)$$

where

$$e_d = \text{driver efficiency}$$
$$khp = \text{wire horsepower, hp}$$

The product of the pump efficiency and driver efficiency is known as the wire-to-water efficiency and is important in calculating the energy cost of pumping water. The wire-to-water efficiency is defined as

$$e_w = e_p e_d \qquad (3.8.4)$$

where e_w = wire-to-water efficiency.

The annual power cost for a pump can be estimated as

$$C = 0.0164 Qh\,PF/e_w \qquad (3.8.5)$$

where

$$C = \text{annual power cost, \$/yr}$$
$$Q = \text{average flow, gpm}$$
$$h = \text{average head, ft}$$
$$P = \text{cost of power, cents/kwhr}$$
$$F = \text{fraction of year pump operates.}$$

A pump operates at more than one point on the pump head curve, so the values of flow, head, and efficiency must only be averages. If the pump operates at significantly different operating points over the course of a year, and accuracy is important, it may be desirable to calculate the energy used at each operating point and fraction of time the pump actually operates at each point to determine cost.

EXAMPLE. When a pump runs by itself, it produces 1000 gpm at 300 ft at an efficiency of 70% and when it runs in parallel with another pump it produces 800 gpm at 250 ft at an efficiency of 75%. The pump runs by itself 30% of the time, in parallel 50% of the time and is down 20%. What is the annual energy cost if the cost of power is $0.12/kwhr? The efficiencies given above are wire-to-water.

$$C = 0.0164(12)[(1000)(300)(0.3)/0.7 + (800)(250)(0.5)/0.75] = \$51,500.$$

Similar calculations must be made for each pump in the pump station and the results can be added.

Pumps should be selected so that they operate efficiently both alone and in parallel with the other pumps in the station. Occasionally, a pump may be very efficient when operated by itself, but when another pump is turned on, it may operate at an inefficient point. This is most likely to occur for pumps with a relatively flat characteristic curve. Pump A in the example problem in Section 3.7 illustrates this situation.

3.9. TIME DEPENDENCE

The flows and pressures in a water distribution system do not remain constant but fluctuate throughout the day. There are two time scales on which these fluctuations occur: (1) daily cycles, in which flows are usually high in the morning, level off during the day, increase dramatically in the evening, and decrease to a minimum after bedtime; and (2) transient fluctuations due to waterhammer surges caused by changing flow velocities. In the latter case, the effects are short-lived, although sometimes dramatic, and will not be covered further in this text. The daily cyclic changes are, however, fairly gradual. This enables the engineer to analyze the system hydraulics using a quasi-steady state approach. This means, that the pressures and flows within the system can be calculated using the steady state energy and continuity equations at one point in time; then knowing these values it is possible to project tank levels and pump operation at the end of the time step (i.e., the time interval considered) and use those values to calculate flows and pressures for the next time step. This type of modeling is known as *extended period simulation*.

To solve most design and operation problems, a steady state analysis of the network corresponding to only a single condition at a time is usually adequate. The engineer can simulate low use, average use, peak use, and fire flows throughout the system in separate runs at a fraction of the cost of an extended period simulation. Selecting and using such models is discussed in more detail in Chapter 4.

For extended period simulation the time steps must be fairly short, otherwise the fluctuations in flow and pressure predicted by the model will lag behind the fluctuations in the real system. This can be overcome by using a predictor-corrector technique in which tank levels during a time step are determined based on preliminary estimates of what they would be at the end of the time step, given the use rate at the beginning of the time step.

The key to extended period simulation is the use of the energy equation (2.2.3) to find tank levels with the continuity equation expressed as:

$$H(t + \Delta t) = H(t) + \frac{\Delta t}{A}(Q) \tag{3.9.1}$$

where

$$H(t) = \text{water level in tank at time } t, \; L$$
$$H(t + \Delta t) = \text{water level in tank at time } t + \Delta t, \; L$$
$$\Delta t = \text{length of time step, } T$$
$$A = \text{tank plan area, } L^2$$
$$Q = \text{net flow into tank, } L^3/T.$$

For tanks in which the plan area is not constant (e.g., a spherical tank), A is a function of H.

The following example problem illustrates how a quasi-steady state model calculates flows and tank levels as compared to a true steady state solution for the same simple problem.

EXAMPLE. Consider two tanks connected by 10,000 ft of 24-in. pipe with a C factor of 130. Both tanks are 30 ft high and have an inside diameter of 40 ft. Tank A has a floor elevation of 450 ft while Tank B has a floor elevation of 400 ft. Initially, Tank A is full and Tank B is empty. Determine the water level in Tank B and the flow between the tanks as a function of time. First use a quasi-steady state solution with time steps of 0.1 and 0.2 hours, then a continuous solution. Ignore minor losses.

There are three equations that must be solved simultaneously, the continuity equation for each of the tanks:

$$\frac{dH_A}{dt} = -\frac{Q}{A}, \quad \frac{dH_B}{dt} = \frac{Q}{A}$$

and the energy equation

$$H_A - H_B = KQ^{1.85}$$

where

$$H_A = \text{elevation in tank A, ft}$$
$$H_B = \text{elevation in tank B, ft}$$
$$Q = \text{flow between the tanks, cfs}$$
$$A = \text{area of tanks, sq.ft}$$
$$K = 4.70 L/D^{4.87} C^{1.85}$$

For the numerical (quasi-steady state) solution solve the energy equation for Q and the continuity equations for the water elevations to yield

$$Q(t) = \frac{[H_A(t) - H_B(t)]^{0.54}}{K}$$

$$H_A(t + \Delta t) = H_A(t) - \frac{Q(t)\Delta t}{A}$$

$$H_B(t + \Delta t) = H_B(t) + \frac{Q(t)\Delta t}{A}.$$

Evaluating K and A gives

$$A = \frac{3.14(40)^2}{4} = 1256 \text{ ft}^2$$

$$K = \frac{4.70(10,000)}{2^{4.87}130^{1.85}} = 0.197.$$

To solve these equations, start with $H_A = 490$ ft and $H_B = 400$ ft at $t = 0$, calculate Q, then substitute Q into the equation for elevation and calculate the water elevation at the end of the time step. Repeatedly carrying out this calculation gives the flow and elevation shown in Fig. 3.7 for time steps of 0.1 and 0.2 hours (360 and 720 sec).

The continuous analytical solution can be achieved by eliminating Q/A from the continuity equations and integrating to give

$$H_B = 890 - H_A.$$

Substituting for H_A in the energy equation and substituting $Q = -A(dH_A/dt)$ gives an equation with H_A as a function of time

$$(2H_A - 90)^{-0.54}dH_A = -\frac{dt}{K^{0.54}A}.$$

Integrating and noting $H_A = 490$ ft when $t = 0$ gives

$$H_A = 0.5\left(\frac{-0.92t}{K^{0.54}A} + 5.81\right)^{2.17} + 890.$$

Substituting for K and A gives the solution for elevation as

$$H_A = 0.5\,(-0.00176t + 7.6)^{2.17} + 890.$$

A similar function for Q can be determined by differentiating the above equation and multiplying by A. A plot of the continuous solution is shown as the smooth curve in Fig. 3.7. It illustrates the fact that extremely small time steps would be required to exactly match the continuous solution. For example, after 2160 sec (36 min), the flow is actually 12.9 cfs but the numerical solution predicts 14.2 cfs (for 360 sec time steps) and 16.1 cfs (for 720 sec time steps).

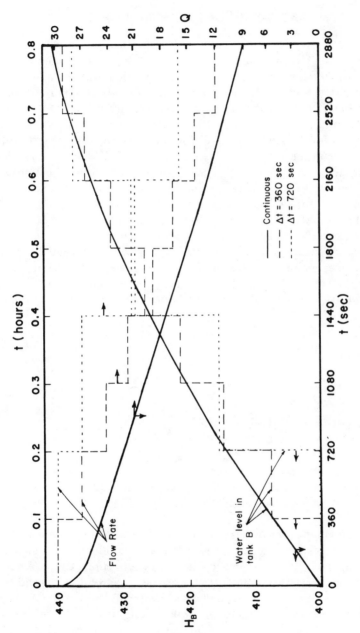

Fig. 3.7. Solution for unsteady flow between two tanks.

3.10. NETWORKS WITH LOOPS OR MULTIPLE CONSTANT HEAD POINTS

Solving a network problem involving loops or multiple constant head points is considerably more difficult than solving a problem with a single source, branched system because the energy and continuity equations must be solved simultaneously. Virtually all water distribution systems are looped systems, so understanding the techniques for solving looped system problems is important for engineers involved with water distribution systems. In this section the overall nature of the solutions is discussed. In the following sections, three techniques for setting up the equations (flow equations, head equations and loop equations) are described. In the following three sections, three commonly used techniques for solving these equations (linear theory method, Hardy-Cross method, and Newton–Raphson method) are described. Readers wishing to learn more about these methods are referred to Jeppson (1976).

The energy equation for flow around a loop was given initially in Eq. (3.4.2). A pipe, or pipes, connecting two constant head points can be regarded as a pseudo-loop in which the head loss between the tanks (i.e., around the pseudo-loop) is dh instead of 0. The general energy equation for flow around a loop is

$$\sum_{i=1}^{N} K_i Q_i |Q_i|^{n-1} - \sum_{j=1}^{M} h_{pj} = dh \qquad (3.10.1)$$

where

K_i = head loss constant in ith pipe
Q_i = flow in ith pipe, L^3/T
N = number of pipes in loop
h_{pj} = head provided by jth pump, L
M = number of pumps in loop
dh = difference in head between constant head points in pseudo-loop (0 in loop), L.

In solving a loop problem, it is essential to define a direction in which flow is considered as positive (e.g., clockwise or, in a pseudo-loop, from higher to lower tank). The $Q|Q|^{n-1}$ term accounts for the direction of flow. Another way of writing this term is $(\text{sgn } Q)|Q|^n$, where sgn Q is $+1$ if $Q > 0$ and -1 if $Q < 0$. (For example, if $Q = -10$ and $n = 2$, $\text{sgn}(-10)\,|-10|^2 = (-1)(100) = -100$.) Both notations for accounting for the direction of flow in the head loss equation will be used in this text. Note that the derivative of $(\text{sgn } Q)|Q|^n$ is $n|Q|^{n-1}$.

There is one loop energy equation for every loop in a network. Sometimes when a network is fairly complicated, it becomes difficult to identify loops

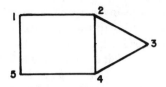

Fig. 3.8. Example network.

without double counting and thus, overspecifying the problem. For example
in Fig. 3.8 there appear to be three loops identified as
  1-2-4-5
  2-3-4
  1-2-3-4-5.
But if the energy equation is written for each of these loops, it can be shown
that one of the energy equations can be derived from the other two. Thus,
there are actually only two independent loops in the network. The rule for
determining the number of independent loops, L, (also called basic loops) is

$$L = P - J + 1 \qquad (3.10.2)$$

where

$$L = \text{number of basic loops}$$
$$P = \text{number of pipes}$$
$$J = \text{number of nodes.}$$

Some examples of this rule are shown in Fig. 3.9.

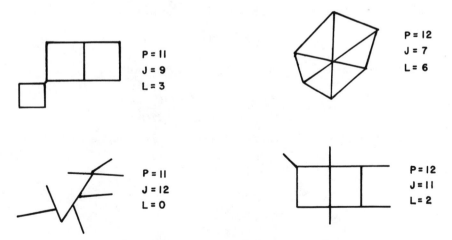

Fig. 3.9. Determining number of independent loops.

The number of pseudo-loops in a network is given by $T - 1$ where T is the number of constant head nodes in the network. (Usually only tanks and reservoirs are treated as constant head nodes.). In cases in which the pump head characteristic curve is very flat or pumps are operating in parallel to yield a roughly constant head over a wide range of flows, it is possible to treat pumps as constant head nodes (instead of using a head characteristic curve) with acceptable error. Similarly, if the pressure upstream of a pressure reducing valve (PRV) is adequate, it is possible to treat the downstream end of the PRV as a constant head node.

EXAMPLE. A classic example of a problem involving pseudo-loops is the three reservoir problem, shown in Fig. 3.10, in which it is necessary to determine the flow in each pipe.
First set up the two energy equations and the continuity equation as

$$0.1 \ (\text{sgn} \ Q_1)|Q_1|^2 + 0.1 \ (\text{sgn} \ Q_3)|Q_3|^2 = 200$$

$$0.08 \ (\text{sgn} \ Q_2)|Q_2|^2 + 0.1 \ (\text{sgn} \ Q_3)|Q_3|^2 = 100$$

$$Q_1 + Q_2 - Q_3 = 20$$

where the convention for signs is shown in the figure. To simplify the problem solve the first energy equation for Q_1 and the continuity equation for Q_2.
Then substitute into the second energy equation and simplify to give the following expression, which is a function of Q_3 only:

$$F(Q_3) = (2000 - Q_3^2)^{0.5} - Q_3 - 20$$
$$+ (1250 - 1.25Q_3^2)^{0.5} = 0.$$

The Newton–Raphson method can be used to solve for Q_3 as follows:

$$Q_3(i + 1) = Q_3(i) - F[Q_3(i)]/F'[Q_3(i)]$$

where i refers to the iteration number. $F'(Q)$ can be expressed as

$$F'(Q_3) = \frac{-Q_3}{(2000 - Q_3^2)^{0.5}} - \frac{1.25Q_3}{(1250 - 1.25Q_3^2)^{0.5}} - 1.$$

Using an initial estimate of 30 cfs, the solution converges rapidly, as shown below:

Q_3	F	F'
30.	−5.72	−5.25
28.9	−0.448	−4.36
28.8	−0.011	

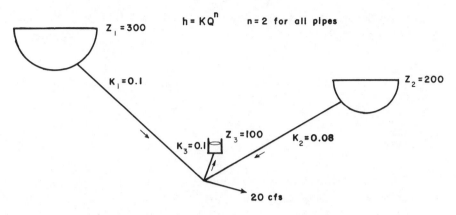

Fig. 3.10. Three-tank example problem.

For $Q_3 = 28.8$ cfs, Q_1 and Q_2 can be calculated as

$$Q_1 = (2000 - Q_3^2)^{0.5} = 34.2 \text{ cfs}$$

$$Q_2 = (1250 - 1.25\,Q_3^2)^{0.5} = 14.6 \text{ cfs}.$$

In simplifying the equations the sgn functions were ignored. In this case the solution was reached, because all of the Q's were positive, but using a different starting value for the iterations (say 0) would have caused the iterative solution technique to fail. For solutions on digital computers, it is best to keep the three equations separate and use a numerical scheme that keeps track of the signs of the flows.

3.11. THE FLOW (Q) EQUATIONS

Solving network flow problems for all but the most trivial situations involves solving a large number of simultaneous, nonlinear equations. There are two steps in solving these equations: setting up the equations so that there are exactly as many independent equations as unknowns, and actually solving the resulting set of equations using some numerical procedure. One approach is to use the flow rates in the various pipes as the unknown and set up a total of P equations (where P is the number of pipes): one energy equation for each independent loop and one continuity equation for each node. Such equations are called the flow (Q) equations because the flows (Q's) are the unknowns. This formulation is discussed in this section. Two alternative formulations, the node (or head) (H) equations and the loop (ΔQ) equations are presented in the following two sections.

The flow equations are given below for a system with L loops (including pseudo-loops) and N nodes.

$$\sum_{i=1}^{m_l} h_{il} - \sum_{k=1}^{P_l} h_{pkl} = dh_l \quad (l = 1, 2, \ldots L) \tag{3.11.1}$$

$$\sum_{i=1}^{n_q} Q_{iq} = U_q \quad (q = 1, 2, \ldots N) \tag{3.11.2}$$

where

h_{il} = head loss in ith pipe in lth loop, L

h_{pkl} = head provided by kth pump in lth loop, L

dh_l = change in head between constant head nodes in lth loop, L

m_l = number of pipes in lth loop

P_l = number of pumps in lth loop

Q_{iq} = flow into qth node from ith pipe connected to the node, L^3/T

U_q = consumptive use at qth node, L^3/T

n_q = number of pipes into qth node.

In order to solve the above equations, it is necessary to substitute the appropriate head loss equation (i.e., Hazen–Williams or Darcy–Weisbach) for the h's and a pump head curve for the h_p's.

The Q equations result in a larger number of equations than the approaches presented in the following sections, but the continuity equations are linear (i.e., Q is in the equation to the first power) and the energy equations can be linearized, so that it is possible to take advantage of some efficient numerical solution methods.

3.12. THE NODE (H) EQUATIONS

The number of equations to be solved can be reduced from $L + J - 1$ to J by combining the energy equation for each pipe with the continuity equation. The head loss equation for a single pipe which previously has been written $h = KQ^n$ can be rewritten as

$$H_i - H_j = K_{ij} |Q_{ij}|^{n_{ij}} \, \text{sgn} \, Q_{ij} \tag{3.12.1}$$

where

H_i = head at ith node, L

K_{ij} = head loss coefficient for pipe from node i to node j

Q_{ij} = flow in pipe from node i to node j, L^3/T

n_{ij} = exponent in head loss equation for pipe from i to j.

(Note that in this section, the notation is different from the convention used in other sections, in that the pipes are designated by the nodes they connect rather than a pipe index number, i.e., pipes have double subscripts.)

Since the head loss is positive in the direction of flow, sgn Q_{ij} = sgn $(H_i - H_j)$ and it is possible to solve Eq. (3.12.1) for Q to give

$$Q_{ij} = \text{sgn}(H_i - H_j)(|H_i - H_j| / K_{ij})^{1/n_{ij}} \qquad (3.12.2)$$

The continuity equation at node i can be written as

$$\sum_{k=1}^{m_i} Q_{ki} = U_i \qquad (3.12.3)$$

where

Q_{ki} = flow into node i from node k, L^3/T
U_i = consumptive use at node i, L^3/T
m_i = number of pipes connected to node i.

The energy and continuity equations can be combined by substituting the energy equation (3.12.2) for each flow in the continuity equation to give

$$\sum_{k=1}^{m_i} \text{sgn}\,(H_k - H_i) \left(\frac{|H_k - H_i|}{K_{ki}}\right)^{1/n_{ki}} = U_i. \qquad (3.12.4)$$

Eq. (3.12.4) is an example of a node (H) equation. There is one such equation for each node, and one unknown (H_i) for each equation. These equations are all nonlinear, so solution techniques different from those used for the Q equations are required.

The node (H) equations are very convenient for systems containing pressure controlled devices (e.g., check valves, pressure reducing valves), since it is a relatively simple matter to fix the pressures at the downstream end of such a valve and reduce the value if the upstream pressure is not sufficient to maintain downstream pressure.

3.13. THE LOOP (ΔQ) EQUATIONS

A third approach to setting up looped system problems is to write the energy equations in such a way that, for an initial solution, the continuity equation will be satisfied. Then it becomes a matter of correcting the flows in each loop in such a way that the continuity equations are not violated. This can be done by adding a correction to the flow to every pipe in the loop. If there is too little head loss, flow is added around the loop; if there is too much, the flow is reduced. The problem thus reduces to one of finding the correction factor ΔQ such that each loop energy equation is satisfied. The loop energy equations may be written

$$F(\Delta Q) = \sum_{i=1}^{m_l} K_i[\text{sgn}\,(Qi_i + \Delta Q_l)]|\,Qi_i + \Delta Q_l|^n = dh_l$$

$$(l = 1, 2, \ldots, L) \qquad\qquad (3.13.1)$$

where

Qi_i = initial estimate of flow in ith pipe, L^3/T
ΔQ_l = correction to flow in lth loop, L^3/T
m_l = number of pipes in lth loop
L = number of loops.

The Qi terms are fixed for each pipe and do not change from one iteration to the next. The ΔQ terms refer to the loop in which the pipe falls. The flow in a pipe is therefore $Qi + \Delta Q$ for a pipe that lies in only one loop. For a pipe that lies in several loops (say, a, b, and c) the flow might be

$$Qi + \Delta Q_a - \Delta Q_b + \Delta Q_c.$$

The negative sign in front of the b term is included merely to illustrate that a given pipe may be situated in the positive direction in one loop and the negative direction in another.

When the loop approach is used, a total of L equations are required, since there are L unknowns, one for each loop.

3.14. NUMERICAL SOLUTION TECHNIQUES

Once the system of equations describing the pipe network has been developed, some numerical technique is required to arrive at a solution. Since most of the equations are nonlinear, the number of techniques which can be used is quite limited. Three commonly used techniques are: (1) the linear theory method, which linearizes the system of equations, solves the linear equations, and substitutes the solution of the linearized equations back into the original system of nonlinear equations to check for convergence; (2) the Newton–Raphson method, which converges to the solution using the derivative of each of the equations to speed convergence; and (3) the Hardy–Cross method, which iterates using one equation at a time, rather than solving a matrix problem, as is done in the other methods.

In general, the linear theory method is most applicable to solving the Q-equations, which are already almost linear. The Newton–Raphson method and the Hardy–Cross method, which can be viewed as a special application of the Newton–Raphson method, are usually applied in the node and loop equation approaches where there are fewer equations, but all of the equations are nonlinear. Examples of each technique and each type of equations are presented in the following sections.

3.15. THE LINEAR THEORY METHOD

Consider the flow equations as given in Section 3.11. The continuity equations are obviously linear. The energy equations are nonlinear, however, since they cannot be written in the form

$$\sum_{i=1}^{N} a_i Q_i = dh \qquad (3.15.1)$$

where

Q_i = flow in ith pipe in loop, L^3/T
a = constant
dh = difference in water elevation in tanks, or 0 for loops, L
N = number of pipes in loop.

The trick to linearizing the energy equations is to let Q_i^{n-1} be a constant for a given iteration. This results in a system of linear equations which can be solved using some of the powerful techniques for solving linear equations (e.g., Gaussian elimination, Jacobi iterative method). The a_i terms are therefore given by

$$a_i = K_i Q_i^{n-1} \qquad (3.15.2)$$

The a_i terms will also contain the coefficients for the pump head characteristic equations. Writing the a terms this way results in a set of $L + N - 1$ linear equations with the same number of unknowns. For this purpose the a's must be double subscripted: the first subscript refers to the number of the loop, while the second refers to the number of the pipe. The entire system of equations becomes

$$\left. \begin{array}{c} a_{11}Q_1 + a_{12}Q_2 + \ldots + a_{1k}Q_k = dh_1 \\ \vdots \\ a_{L1}Q_1 + a_{L2}Q_2 + \ldots + a_{Lk}Q_k = dh_k \end{array} \right\} \quad L \text{ equations}$$

$$\left. \begin{array}{c} b_{11}Q_1 + b_{12}Q_2 + \ldots + b_{1k}Q_k = U_1 \\ \vdots \\ b_{N-1,1}Q_1 + b_{N-1,2}Q_2 + \ldots + b_{N-1,3}Q_3 = U_{N-1} \end{array} \right\} \quad N-1 \text{ equations}$$

where

$$a_{ij} = \begin{cases} K_j|Q_j|^{n-1}, & \text{if pipe } j \text{ is in loop } i \\ 0, & \text{if pipe } j \text{ is not in loop } i \end{cases}$$

$$b_{ij} = \begin{cases} +1, & \text{if flow in pipe } j \text{ is positive into node } i \\ -1, & \text{if flow in pipe } j \text{ is negative into node } i \\ 0, & \text{if pipe } j \text{ is not connected to node } i. \end{cases}$$

To solve the network problem it is necessary to solve the linear system of equations, recalculate the a terms, and resolve the linear equations repeatedly until the solution converges. The linear theory method tends to overcorrect the Q_i's so that it is possible to base the a terms, not on the new value of the Q_i's but on a weighted average of the old and new Q_i's. This tends to speed convergence. One advantage of the linear theory method is that it does not require an initial solution. Therefore all of the Q_i's can be set to some arbitrary value and the method will still produce an answer. Of course, a good initial solution will result in quicker convergence. The best way to understand the linear theory method is to follow the example problem below.

EXAMPLE. Consider the simple looped network shown in Fig. 3.11. Find the flows in each pipe.

There is one loop energy equation, plus two independent continuity equations. Consider the flow to be positive in the clockwise direction, so that these equations become

$$(1.)|Q_1|^{0.85}Q_1 + 0.1|Q_2|^{0.85}Q_2 + 1.5|Q_3|^{0.85}Q_3 = 0$$
$$(1.)Q_1 + (-1)Q_2 \qquad\qquad = 2.$$
$$(1.)Q_2 + (-1.)Q_3 \quad = 1.5.$$

This problem is simple enough that the solution to the linear set of equations can be written analytically as follows:

$$Q_1 = \frac{2a_2 + 3.5a_3}{a_1 + a_2 + a_3}$$

$$Q_2 = Q_1 - 2.$$

$$Q_3 = Q_2 - 1.5 = Q_1 - 3.5$$

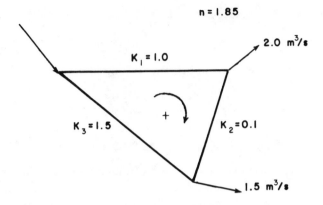

Fig. 3.11. Network for linear theory example.

where

$$a_1 = |Q_1|^{0.85}$$
$$a_2 = 0.1|Q_2|^{0.85}$$
$$a_3 = 1.5|Q_3|^{0.85}.$$

Instead of using the new Q's as the basis for the a terms in the next iteration, the Q's in the next iteration will be calculated from a weighted average of old and new Q's given by

$$Q(\text{next}) = [2Q(\text{new}) + Q(\text{old})]/3.$$

The solution is shown below in tabular form starting with an initial solution in which all of the Q's are 1.

Next						New		
Q_1	Q_2	Q_3	a_1	a_2	a_3	Q_1	Q_2	Q_3
−1.00	1.00	1.00	1.0	0.01	1.5	2.09	0.09	−1.40
1.72	0.39	−0.60	1.6	0.05	0.97	1.34	−0.66	−2.16
1.47	−0.31	−1.64	1.4	0.04	2.3	2.17	0.17	−1.32
1.94	0.01	−1.43	1.8	0.00	2.0	1.88	−0.12	−1.62
1.90	−0.08	−1.56	1.7	0.01	2.2	1.95	−0.05	−1.55

Thus, the solution

$$Q_1 = 1.93, \qquad Q_2 = -0.06, \qquad Q_3 = -1.55.$$

The head losses may be checked as follows:

$$|1.93|^{0.85}1.93 + 0.1|-0.06|^{0.85}(-0.06) + 1.5|-1.55|^{0.85}(-1.55) = 0.00.$$

Continuity at the nodes can also be used as a check:

$$1.93 - (-0.06) = 1.99$$
$$-0.06 - (-1.55) = 1.49.$$

3.16. THE NEWTON–RAPHSON METHOD

The Newton–Raphson method is a powerful numerical method for solving systems of nonlinear equations. It can be derived from the definition of a derivative. The Newton–Raphson method is suited for problems that can be expressed as $F(x) = 0$, where the solution is the value of x that will force F to be zero. The derivative of F can be approximated by

$$\frac{dF}{dx} = \frac{F(x + \Delta x) - F(x)}{\Delta x}. \tag{3.16.1}$$

Given an initial estimate of x, the solution to the problem is the value of $x + \Delta x$ that forces F to 0. Setting $F(x + \Delta x)$ to zero and solving for Δx gives

$$\Delta x = -F(x)/F'(x) \tag{3.16.2}$$

and the new value of $x + \Delta x$ becomes x for the next iteration. This process is continued until F is sufficiently close to zero.

The discussion above is relevant when there is only one equation with one unknown. However, in pipe network problems there are many equations with many unknowns. The Newton–Raphson method, as it can be applied to the $N - 1 = k$ H-equations is described in the following paragraphs although the method also is applicable to the ΔQ equations.

For each node (1 through k), it is possible to write a head equation of the form

$$F(H_i) = \sum_{j=1}^{m_i} [\text{sgn}\,(H_j - H_i)] \left(\frac{|H_j - H_i|}{K_{ji}}\right)^{1/n_{ij}} - U_i = 0$$

$$(i = 1, 2, \ldots, K) \tag{3.16.3}$$

where

$$m_i = \text{number of pipes connected to node } i$$
$$U_i = \text{consumptive use at node } i, \ L^3/T.$$

If the value of one of the F's at the ith iteration is $F(i)$, then the difference between the ith and $(i + 1)$th iteration is

$$dF = F(i + 1) - F(i) \qquad (3.16.4)$$

This change can also be approximated by the total derivative

$$dF = \frac{\partial F}{\partial H_1} \Delta H_1 + \frac{\partial F}{\partial H_2} \Delta H_2 + \ldots + \frac{\partial F}{\partial H_k} \Delta H_k \qquad (3.16.5)$$

where ΔH = change in H between the ith and $(i + 1)$th iterations, L. The problem reduces to one of finding the values of ΔH which force $F(i + 1) = 0$. This may be done by setting Eq. (3.16.4) and (3.16.5) equal. This results in a system of k linear equations with k unknowns (ΔH) which can be solved by the same linear methods described earlier.

The solution procedure consists of picking some initial values of H, calculating the partial derivatives of each F with respect to each H, solving the resulting system of linear equations to find H, and repeating until all of the F's are sufficiently close to 0.

The key step in solving problems of this type is remembering that the derivative of the terms in Eq. (3.16.3) can be given by

$$\frac{d}{dH_j} [\text{sgn } (H_i - H_j)] \left(\frac{|H_i - H_j|}{K_{ij}} \right)^{1/n_{ij}} =$$

$$\frac{-1}{(n_{ij})(K_{ij})^{1/n_{ij}}} (H_i - H_j)^{(1/n_{ij}) - 1}$$

and

$$\frac{d}{dH_i} [\text{sgn } (H_i - H_j)] \left(\frac{|H_i - H_j|}{K_{ij}} \right)^{1/n_{ij}} =$$

$$\frac{1}{(n_{ij})(K_{ij})^{1/n_{ij}}} (H_i - H_j)^{(1/n_{ij}) - 1} \qquad (3.16.6)$$

EXAMPLE. Consider the network shown in Fig. 3.12. Find the flow in each pipe and the head at node 2.

The H-equation for the two pipe system (only one equation is required since there is only one node with an unknown head) is given by

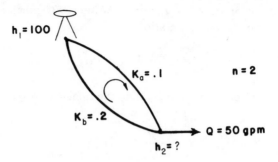

Fig. 3.12. Network for node equation example.

$$F(H) = [\text{sgn} \,(100 - H)] \left(\frac{100 - H}{0.1} \right)^{0.5}$$

$$+ [\text{sgn} \,(100 - H)] \left(\frac{100 - H}{0.2} \right)^{0.5} - 50 = 0$$

Only one derivative is required

$$\frac{dF}{dH} = -\frac{10}{2} \left(\frac{100 - H}{0.1} \right)^{-0.5} - \frac{5}{2} \left(\frac{100 - H}{0.2} \right)^{-0.5}$$

The new value of H will be calculated from the previous value as

$$H(i + 1) = H(i) - F/(dF/dH).$$

For an initial estimate of H of 50 ft, the solution is shown below:

H	F	dF/dH
50.00	−11.83	−0.382
19.11	1.43	−0.300
14.27	0.02	−0.292
14.20	0.00	−0.197
15.15	−0.27	−0.201
13.97	0.12	−0.199
14.41	−0.06	−0.200
14.12	0.03	−0.200
14.25	−0.01	

Knowing $H = 14.25$ ft, the flows can be calculated as

$$Q_1 = [\text{sgn }(100 - 14.25)]\left(\frac{100 - 14.20}{0.1}\right)^{0.5}$$

$$= 29.29 \text{ gpm}$$

$$Q_2 = [\text{sgn }(100 - 14.20)]\left(\frac{100 - 14.20}{0.2}\right)^{0.5}$$

$$= 20.71 \text{ gpm.}$$

To check:

$$29.29 + 20.70 = 49.99 \text{ gpm.}$$

3.17. THE HARDY-CROSS METHOD

By solving the system of governing equations simultaneously, the linear theory method and the Newton–Raphson method can converge to the correct solution rapidly. For anything but an extremely simple case, the equations must be solved using a computer. Manual solutions or solutions on small computers may not be possible with these methods. However, the Hardy–Cross method, which dates back to 1936, can be used for such calculations. In essence, the Hardy–Cross method is similar to applying the Newton–Raphson method to one equation at a time.

The Hardy–Cross method is usually applied to the ΔQ equations although it can be applied to the node equations and even the flow equations. Convergence problems have been reported for the Hardy–Cross method, especially in systems with pumps or check valves. The method, when applied to the ΔQ equations, requires an initial solution which satisfies the continuity equation. Nevertheless it is still widely used, especially for manual solutions and small computers or hand calculators, and produces adequate results for most problems.

For the lth loop in a pipe network the ΔQ equation can be written as follows:

$$F(\Delta Q_l) = \sum_{i=1}^{m_l} K_i[\text{sgn}(Qi_i + \Delta Q_l)]\,|Qi_i + \Delta Q_l|^n - dh_l = 0 \quad (3.17.1)$$

where
  ΔQ_l = correction to lth loop to achieve convergence, L^3/T
  Qi_i = initial estimate of flow in ith pipe (satisfies continuity), L^3/T.
  m_l = number of pipes in loop l

Applying the Newton–Raphson method for a single equation gives

$$\Delta Q(k + 1) = \Delta Q - \frac{\sum_{i=1}^{m_l} K_i(Qi_i + \Delta Q_i)|Qi_i + \Delta Q_l|^{n-1}}{\sum_{i=1}^{m_l} K_i n_i |Qi_i + \Delta Q_l|^{n-1}} \qquad (3.17.2)$$

where the $k + 1$ refers to the values of ΔQ in the $(k + 1)$th iteration. All other values refer to the kth iterations and are omitted from the equation for ease of reading. Eq. (3.17.2) should be recognized as being equivalent to

$$\Delta Q(k + 1) = \Delta Q(k) - F(k)/F'(k). \qquad (3.17.3)$$

It must be remembered that the signs on the Qi terms depend on how that pipe is situated in the loop under consideration. The same pipe may have different signs in different loops.

EXAMPLE. Consider the pseudo-loop system shown in Fig. 3.13. Find the flows in each pipe and the head at each node.
    Let the initial solution be

$$Qi_1 = 5. \text{ cfs}$$
$$Qi_2 = 3. \text{ cfs}$$
$$Qi_3 = 0. \text{ cfs}$$
$$Qi_4 = -5. \text{ cfs.}$$

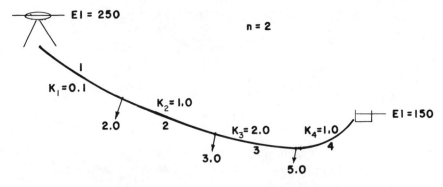

Fig. 3.13. Network for loop equation example.

The equation to be solved is

$$F(\Delta Q) = 0.1|5 + \Delta Q|(5 + \Delta Q) + 1.0|3 + \Delta Q|(3 + \Delta Q)$$
$$+ 2.0|\Delta Q|(\Delta Q) + 1.0|-5 + \Delta Q|(-5 + \Delta Q) - 100$$
$$= 0.$$

The derivative is

$$F'(\Delta Q) = 0.2|5. + \Delta Q| + 2.|3. + \Delta Q| + 4.|\Delta Q| + 2.|-5. + \Delta Q|.$$

The formula for generating a value of ΔQ from previous values is

$$\Delta Q(k + 1) = \Delta Q(k) - F(k)/F'(k)$$

where k is the iteration number. For the first iteration ΔQ is zero and the following solution can be generated:

ΔQ	F	F'
0.00	−113.5	12.0
9.46	274.9	70.1
5.54	45.6	41.9
4.45	3.7	35.1
4.34	−0.0	34.6
4.34		

With the solution for ΔQ it is now possible to calculate the individual flows as $Q = Qi + \Delta Q$:

$$Q_1 = 9.34 \text{ cfs}$$

$$Q_2 = 7.34 \text{ cfs}$$

$$Q_3 = 4.34 \text{ cfs}$$

$$Q_4 = -0.66 \text{ cfs.}$$

The heads can be calculated at each node as follows:

$$H_{1,2} = 250 - 8.72 = 241.3 \text{ ft}$$

$$H_{2,3} = 241.3 - 53.9 = 187.4 \text{ ft}$$

$$H_{3,4} = 187.4 - 37.7 = 149.7 \text{ ft.}$$

REVIEW QUESTIONS

1. What is the relationship between pump efficiency, driver efficiency and wire-to-water efficiency?

2. What type of equation is usually used to approximate a pump head characteristic equation?

3. If you are going to buy a pipe network model to use on your home computer, what kind will you probably buy? Why?

4. What are the primary differences between the Hardy–Cross and Newton–Raphson method for solving the ΔQ equations?

5. The linear theory method is more applicable to which formulation of the pipe network equations?

6. The operating point of a pump can be found at the intersection of what two curves?

7. How many ΔQ equations must be set up for a network with L loops (and pseudo-loops), N nodes, and P pipes? How many H-equations must be set up?

8. For two pipes in parallel, with $K_1 > K_2$, what is the relationship between K_1, K_2, and K_e, the K for the equivalent pipe replacing 1 and 2 ($h = KQ^n$)?
   a. $K_1 > K_2 > K_e$
   b. $K_1 > K_e > K_2$
   c. $K_e > K_1 > K_2$

9. Why is it significantly more difficult to determine flows and heads in a branched system with two tanks as opposed to a system with one tank?

10. In fitting a pump head curve with an equation of the form $h_p = AQ^2 + BQ + C$, what is the sign of A and why?

11. Given two pipes that are connected in parallel, what is the relationship between flows and head in the two pipes?

12. In a pumping station, are pumps more likely to be connected in parallel or series? Why?

13. What kinds of problems require an extended period simulation?

PROBLEMS

1. Consider the two tank system shown in Fig. 3.1p. For water use rates U of 2, 7, and 10 cfs, find the flow rates in pipes 1 and 2. Draw a graph showing Q_1 and Q_2 versus U.

2. Use the Newton–Raphson method to determine the flow in each line of the

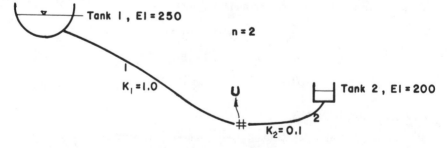

Fig. 3.1p. Network for Problem 1.

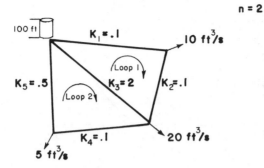

Fig. 3.2p. Network for Problem 2.

network shown in Fig. 3.2p. For an initial solution use, $Q_1 = 10$, $Q_2 = 0$, $Q_3 = -20$ (+20 for loop 2), $Q_4 = 0$, and $Q_5 = -5$.

(*Solutions*: 20.12, 10.12, −+5.04, −4.84, and −9.84 cfs.)

3. If a tank contains volume V at time $t = 0$, what is the volume at time t if the outflow during that time is Q?

4. Prove that

$$\frac{d[\operatorname{sgn} f(x)] f(x)^n}{dx} = nf(x)^{n-1} \frac{d[f(x)]}{dx}.$$

5. For the system head curves shown in Fig. 3.3p, which is for an uphill (lift) system? For the other system, what is the flow when the pumps are off?

6. Given the pump head characteristic curve in Fig. 3.4p, draw in the curve for two identical pumps operating in parallel.

7. Consider three 1000-ft-long pipes in parallel, each with a Hazen–Williams C factor of 120. Their diameters are 24, 16, and 12 in. respectively. What is the C factor of a 1000-ft-long 24-in. pipe that is equivalent to these three pipes in parallel? (*Solution*: 181)

8. Replace two identical pipes ($C = 130$) with a single equivalent pipe with the same diameter and length. What is the C factor of that pipe? Roughly what is the magnitude of the roughness for this equivalent pipe?

9. Using the H-equations, find the pressure at A (H_A) when the flow at A is 10 cfs, for the system shown in Fig. 3.5p.

10. (a). Given the system shown in Fig. 3.6p, find the equivalent K for the piping system using $h = KQ^{1.85}$. (Ans. 0.1944)

(b) Develop a system head curve for the pipes for flows from 0 to 30 cfs.

(c) Given pump characteristic data for pumps A and B in the accompanying table, find the operating point when

(1) A

(2) B

(3) A and B are operating. (*Solution*: (1) 15 cfs, 212 ft; (2) 23.5 cfs, 255 ft; (3) 28.5 cfs., 280 ft)

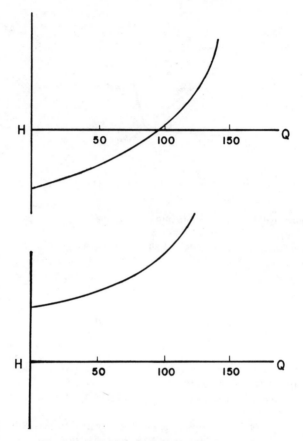

Fig. 3.3p. System head curves for Problem 5.

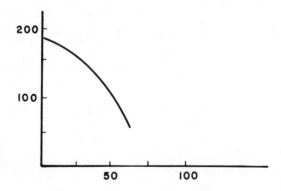

Fig. 3.4p. Pump head curves for Problem 6.

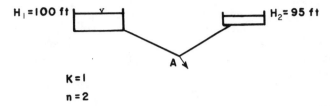

Fig. 3.5p. Network for Problem 9.

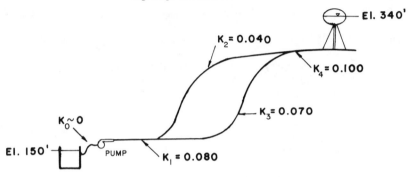

Fig. 3.6p. Network for Problem 10.

Pump Characteristics Data for Problem 10

Pump A

Q (cfs)	Efficiency (%)	Head (ft)	BHP (hp)
0	0	320.	270.
5	52.6	295.	318.
10	78.0	269.	391.
12	79.7	251.	429.
15	74.6	212.	486.
20	44.0	112.	582.

Pump B

Q (cfs)	Efficiency (%)	Head (ft)	BHP (hp)
0	0	320.	
5	35.	310.	500.
10	60.	305.	577.
20	80.	280.	794.
25	75.	245.	927.
30	60.	200.	1135.

(d) Given that the pumps and drivers have an average wire-to-water efficiency of 55%, and the price of energy is 5 cents per kilowatt-hour, calculate the annual average energy costs if pump A operates alone 30% of the time, pump B operates alone 40% of the time and pumps A and B operate together 30% of the time? (*Solution:* $384,000/yr)

REFERENCES

Hicks, T.G. and T.W. Edwards, 1971, *Pump Applications Engineering,* McGraw-Hill, New York.

Jeppson, R.W., 1976, *Analysis of Flow in Pipe Networks,* Ann Arbor Scientific Publishers, Ann Arbor.

Karassik, I.J., et al, 1976, *Pump Handbook,* McGraw-Hill, New York.

Sanks, R.L. (ed.), 1981, Proc. Conf. Pump Station Design for the Practicing Engineer, Montana State Univ., Bozeman, MT.

4 | USING WATER DISTRIBUTION SYSTEM MODELS

4.1. INTRODUCTION

In Chapters 2 and 3 the theory of flow in pipes was described and methods for solving complex pipe network problems were presented. While it is important to understand the theory for solving these problems, the engineer must also accomplish the difficult task of linking the theory with the physical problem. At present, the complexity of real systems exceeds the ability of engineers to model every valve and bend and consider every possible operating condition. Thus a critical part of analyzing a water distribution system is how to combine the numerical techniques embodied in a computer model with an often sketchy description of the physical system to arrive at a model of the system that can be used with confidence.

When discussing mathematical models of the hydraulics of water distribution systems, most individuals use the term *model* to refer only to the numerical method used to solve the network equations. However, in reality a model of a specific system consists not only of a computer program but also of data describing the system. Usually, these data are the weakest link in the modeling process, and because of this many engineers end up playing a game of GIGO (Garbage In, Garbage Out). Computer programs can produce results that are accurate to several decimal places, but when the data used to drive the model are only accurate to ±20%, the results are only as accurate as the least accurate datum.

Developing a model of a distribution system is significantly different from writing a program to solve for flows in a pipe network. In Chapter 3, it was always assumed that the pipe characteristics and water use distribution were known. In this chapter, methods for determining water use and pipe characteristics are discussed along with the problems of how to manage all the data involved in analyzing water distribution systems. Selecting the approach and the computer program to be used is the first topic to be presented. The question of how to condense large real systems to the type of pipe network a

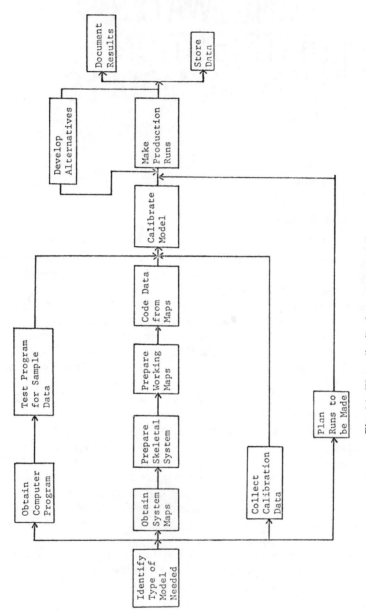

Fig. 4.1. Water distribution system modeling process.

computer program can handle with acceptable accuracy and cost is addressed. Finally, model calibration is considered.

The data entered into the model are usually not known precisely, so model calibration is critical if the model is to truly represent the actual system. Only when the model is satisfactorily calibrated can it be used to study the system.

The steps involved in developing and using a water distribution model are shown in Fig. 4.1.

4.2. SELECTING THE APPROACH AND MODEL

Virtually any engineer with rudimentary skill in computer programming can write a program to solve a pipe network problem. Readers of this text may have used a computer to solve some of the problems at the end of the previous chapter. However, there is a significant difference between solving simple problems and modeling real systems. Fortunately, the engineer does not need to invest the enormous amount of resources required to write, debug, and document a computer program because many excellent programs are already available. Some are free to the user, some can be purchased at nominal cost, and still others can be leased from engineering or time-sharing firms. Most commercially available programs are fairly machine independent, so they can be transported easily from one computer to another.

In selecting a computer program, one should keep several considerations in mind. These include:

1. Steady state versus extended period,
2. Simulation versus optimization,
3. Ease of making the program operational,
4. User-friendliness with respect to input and output,
5. Need for loop identification or initial solution from user,
6. Numerical method used for solution,
7. Method for accounting for pumps and valves (especially pressure reducing and check valves),
8. Head loss equation used,
9. Ability to interface with computer graphics.

Each of these considerations is discussed in more detail below.

Traditionally, water distribution models have been used to simulate flow and pressures in a given system for a single point in time. This is all that is required for most problems, but there are some instances where it is desirable, and even necessary, to model the behavior of a system over an extended period of time. Examples of such problems include determining how long it will take tanks to drain or fill or when pumps should be operated. Extended period simulations are not a true non-steady state simulation, but consist of a steady state model run with time-varying boundary conditions (i.e. tank water levels, water use distributions) which were projected by the results of the steady state model at the previous time step. For most studies a steady state model is adequate and considerably less expensive to run.

Most currently available water distribution models are simulation models in that, for a given system and water use, they predict flows, pressures, and, in the case of extended period models, tank elevations. However, some models exist which solve the problem in the other direction (i.e., given flows and pressure in the system, they determine the pipe sizes, pump capacities, etc. for the least cost system). These models are referred to as optimization models, since they determine optimal conditions rather than merely predict what will occur. Optimization models require considerable extra information on the costs of alternatives, and constraints to be met, and are only now available as user oriented products, as they are still somewhat restrictive in the way they model systems and require considerable computer time. Thus, at present, simulation models are almost universally used with a trial-and-error approach for pipe sized selection for real systems. Optimal pipe size selection is discussed in more detail in Chapter 5.

In selecting a program, the engineer should make certain that the user's guide and documentation are both available and understandable. A well documented model will very seldom cause problems, but models without documentation are immediately questionable since the user cannot be sure the model developer understands all the nuances of the program. It is generally best to run a program on the computer system on which it was developed. If the user must move the program to a different system, the move will be easiest if the selected program is written in a standard language such as Fortran 77. A few nonstandard statements in a program can give a user fits when converting to a new system.

Most programs require input data in fixed columns on data cards which must be in a specified order. Others are more forgiving, allowing the user the convenience of keyword-oriented free-format input. This spares the user inconvenience and lost computer time if, for example, a decimal point is left off a number.

A good program should have plenty of clear diagnostic messages to tell a user why a run was unsuccessful or why some output is in question. Messages such as "A7 error on J card—stop at 077787" are not very helpful to the typical user. Good diagnostics should indicate the cause of the problem and the user's response required to solve it.

Similarly, the user should have control over the level of detail of the output. In addition to echoing the user's input and printing the flow in each pipe and pressure at each node, the program should print such information as energy grade line slope, head and flow provided by pumps, and the setting of pressure reducing valves. It should also be able to provide information on the number of iterations required and the final value of the convergence indicator (usually largest correction factor or loop with the largest absolute value for head loss). Programs should be able to print solutions after a few iterations and then continue so that it is possible to test the sensitivity of solutions to the number of iterations. The user should also be able to determine the cost of each run so that intelligent decisions can be made on the tradeoffs between detail, accuracy, and cost. This should prevent the user from spending the entire computer budget on the first three runs.

Some of the less sophisticated programs require the user to identify the loops in the system or provide an initial solution that satisfies the continuity equations. For all but the simplest problems this is a tedious procedure in which it is easy to make errors. Better programs are self-starting in that they do not need initial solutions. Better programs can also resolve the topology of the system, determining which pipes are in which loops and keeping track of direction of flow in each loop.

The numerical method which is actually used to solve the pipe network problem is the part of the program that probably received the most attention from the developer when the program was written. Any of the standard techniques described in the previous chapter will be adequate for most problems. In selecting a method, the engineer should remember that the Hardy–Cross method operates on one loop (or node) at a time and, thus, can be used easily on small computers. However, this asset of the Hardy–Cross method is also a liability in that, since the system of equations is not considered in matrix form, convergence is somewhat slower than the linear theory or Newton–Raphson methods. In general, any tested program using any of the standard solution techniques or variations thereof, will be adequate for most problems. Collins (1980) has shown, however, that convergence to the correct solution may be extremely slow for some cases and there may not be unique solutions in others. Even so, ease of use of the program is a more important criteria than number of iterations required to solve a benchmark problem. A simple program with plenty of "whistles and bells" is more useful than an efficient program that can only be run by its developer. Comparisons between different programs should be based on actual cost to run the programs rather than number of iterations, since an iteration with one of the matrix techniques uses a good deal more computer time than an iteration with the Hardy–Cross method.

Flexibility is more important than efficiency in selecting a program. This means that a program should not only work for the garden variety networks used as examples in Chapter 3 but for the type of network which will be encountered in the real world. These networks include in-line booster pumps, hydropneumatic tanks, check valves, and pressure reducing valves, all of which make convergence slower because they allow flow in only one direction and operate automatically, so that a model user does not usually know what state the valve or pump will be in before making a run. The ease with which a model handles these special features is an important criterion in program selection.

As discussed in Chapter 2, head loss in individual pipes is usually determined using the Hazen–Williams or Darcy–Weisbach equation. While the Darcy–Weisbach equation is more theoretically correct and can be applied to virtually any fluid, the Hazen–Williams equation does not require reevaluation of the friction factor at every iteration. Since engineers have a good feel for the value of the Hazen–Williams C factor over a wide range of pipe sizes, types, and ages, the Hazen–Williams equation can be used with acceptable error, especially for hydraulically smooth pipes. Some of the better programs give the user the option of selecting either equation, or correcting the C factors for high velocities in rough pipes.

With improvements in both software and hardware for computer graphics, it is now possible to develop plots of the network showing flows and pressure directly from the computer. This makes it much easier for the user to visualize what is happening in the network. Purchasing the hardware and software and providing the training for personnel to take advantage of this capability is costly, but is likely to be justifiable if the required resources are available in-house, the programs will be used often and the networks to be studied are complicated.

Once a computer program for solving network problems has been obtained, it should first be tested using data from an example problem in the user's guide. If the test run results do not agree with the results of the example, the program developer should be contacted to determine the cause of the discrepancy once the accuracy of the input data has been checked.

4.3. SYSTEM MAPS

Even while the selected program is being tested, the engineer should obtain maps of the water distribution system and prepare a working map of the pipe network to be modeled. Selecting the scale of the working map is critical since it is very desirable for the maps to fit in one sheet, even if the sheet is very large. The maps, though, should not be so cluttered that it is impossible to clearly number nodes and pipes. Maps on the scale of 1 : 24,000 are useful for modeling skeletal systems, but such maps will require the use of magnifying glasses if the model is to be used to model within neighborhood systems. A 1 : 10,000 scale is better for this level of detail. As-built drawings on a 1 : 2400, or smaller, scale require too many sheets to represent the entire system, so that matching pipes from one sheet to the next quickly becomes a nightmare. As-built drawings can, however, be reduced and combined into a mosaic of the network, since they contain a great deal of valuable information if they have been properly prepared. The ideal base map contains street names and elevations or contour lines, preferably accurate to ± 1 ft.

In some cases there are discrepancies between different maps of a system because there may have been errors in drawing the map or the maps may be based on plans that were never implemented. As-built drawings, contract documents, and conversations with utility personnel should be used to resolve any inconsistencies.

Usually it is not necessary to include every pipe in the distribution system in the model. Small pipes, especially those perpendicular to the usual direction of flow, can be eliminated to yield a skeletal system which behaves like the actual system but is much easier to work with and saves considerable computer time. Eggener and Polkowski (1976) studied the effect of skeletonizing on the results produced by models and found that the resulting errors are small if the skeletonization is done properly. Elimination of small pipes carrying large amounts of water (e.g., near the source or large users) produced the largest errors, so elimination of pipes simply based on their diameter may not be satisfactory. The pipes to be eliminated from the model should be limited to those carrying an insignificant portion of the flow. Usually if a pipe has a diameter less than half that of the other pipes in a given portion of the system, it can be eliminated with acceptable error.

The section in Chapter 3 on equivalent pipes gives some rules for deciding if a pipe can simply be ignored or if it should be accounted for by including its carrying capacity in a large, nearby main. The approach to skeletonizing the system also depends on the type of network equations to be solved. If the node equations are being solved the user should try to reduce the number of nodes; if the loop equations are being solved the user should try to reduce the number of loops; and if the pipe (flow) equations are being solved, the user should try to reduce the number of pipes.

The resulting skeletal working maps of the system are used to develop data for the model and also provide a way of keeping track of alternatives. If the maps are stored on the computer, the program will automatically produce maps of each network. If the maps are drawn by hand, they should be drawn on paper from which blue line copies can be made. In this way it is possible to draw the results of each important model run on the base map, so that the data can be saved for comparison of alternatives later in the study.

Color coding of data on the base map also makes data entry easier. For example, label pipes in black, nodes in blue, elevations in red, water use in brown, etc. While this may seem fairly trivial, it spares the engineer from questions such as, "Was node 120 at elevation 150 or is it the other way around?" Data written on the map should be expressed in the same units as required by the program. If the program wants water use data in gallons per minute, label the map in those units, not cubic feet per month.

4.4. INITIAL ESTIMATES OF PIPE ROUGHNESS AND WATER USE

Contrary to the impression that may have been given in the example problems in Chapter 3, the engineer does not know the precise values of internal pipe roughness (which will be referred to as C factor in this discussion) and the water use in the system to be simulated. Adjusting the C factors and water use so that model predictions agree with observed pressures and flows is known as *calibration*. The calibration process is made much easier if the initial values of C and use are fairly accurate values. In such cases, calibration consists of only some minor adjustments to the input data.

Initial C values can be found in the literature. A good source is the original book on the subject by Williams and Hazen (1920). Table 2.3 gives good typical values for C. These values can be converted into pipe roughness using Fig. 2.8 for those who prefer the Darcy–Weisbach equation. The engineer must remember that the values in the tables are merely typical values, which may need to be adjusted during the calibration process.

Assigning water use rates to each of the nodes in a system is a difficult process subject to considerable error because unlike the C factor, which usually varies gradually over a span of decades, water use rates change from minute to minute. (How can an engineer know who was flushing the toilet when pressure readings for model calibration were collected?) To make the problem more difficult, the engineer must actually prepare several sets of water use estimates: one that reflects the conditions during model calibration, plus additional sets which reflect water use projected over the planning horizon. The latter are often estimated by multiplying the water use for calibration times a constant to account for growth (or conservation). This is, however, not necessarily the best approach. Use estimates for future years should be based on carefully thought out scenarios of industrial and residential development. All of the growth may occur in one portion of town or the growth may be distributed evenly around the system or among a small number of areas.

Methods for predicting water use were developed by Linaweaver, Geyer, and Wolff (1966) as part of the Residential Water Use Research Project at The Johns Hopkins University. They have been summarized succinctly by Clark, Viessman, and Hammer (1977) and Boland, Bauman, and Dziegielewski (1981). Additional data are available from AWWA Manual M22 (1975). These methods should be used for projecting future use, rather than simply multiplying current use by a growth (conservation) factor.

There are two approaches to developing water use distributions (also called loading distribution). The first approach involves aggregating individual users while the second involves disaggregating the system wide use. In aggregating individual water users, first assign large water users to nodes. This would include industries and large commercial establishments. Then calculate an average water use at homes. This may vary with the neighborhood. Then multiply the number of homes near a node by this average use and add the product to the large individual users to obtain use at each node. The sum of the use at the individual nodes may fall short of the production of water at the treatment plant(s). This difference is the unaccounted—for water, and should be spread evenly through the system, unless the engineer has reason to believe it is confined to a given area.

The disaggregated approach starts with the average day water use in the season under consideration. Assign large commercial and industrial users to nodes, then spread the remaining water use through the system on the basis of gpm/node weighted by the judgment of the engineer. The second method is not as accurate as the first but is much easier. Accuracy can be greatly improved if master meter readings from individual neighborhoods, or water audit type results, are available.

Water use does not remain constant but fluctuates in seasonal and diurnal cycles. Utilities usually only have data on how average daily water use fluctuates through their system. The estimates of water use described in the previous paragraphs must be corrected to account for the water use at the time the model is being calibrated. For small systems with little storage and few industries that operate around the clock, the ratio of peak to average water use rates can be as high as 6:1, while in large systems with considerable storage and 24-hour industries this ratio can be more on the order of 2:1. For calibration purposes, the water use during the day can be taken initially as 1.5 times the average use. This number can then be adjusted during calibration. Some engineers have been known to collect data for model calibration in the middle of the night in order to achieve almost complete control of flows in the system by opening hydrants.

4.5. CALIBRATION OVERVIEW

Model calibration consists of adjusting model input data (usually initial estimates of C and water use) so that the model accurately predicts flows and pressures observed in the system. In some instances in which the data for calibration are not very precise, it is possible for a model to appear to be calibrated simply because errors in C, tank levels, pump operation and water use compensate for one another.

Calibration of a water distribution model is a two-step process consisting of: (1) comparison of pressures and flows predicted by the model with observed pressures and flows for known operating conditions (i.e., pump operation, tank levels, pressure reducing valve settings, etc.), and (2) adjustment of the model input data to improve agreement between predicted and observed values. In the case of an extended period simulation, the model should also be able to predict tank water levels and pump and valve operation. A model can be considered calibrated for a given set of boundary, or initial, conditions (depending on whether it is a steady state or extended period simulation) if it can predict flows, pressures, and tank levels with reasonable agreement. Calibration for one set of operating conditions does not imply calibration in general, although confidence in the accuracy of results from the model should increase with an increase in the range of conditions for which the model is calibrated.

Quantifying what is meant by "reasonable agreement" is difficult because it depends on: (1) the quality of the pressure and elevation data used, and (2) the amount of effort the model user is willing to spend fine tuning the model. While a good-quality pressure gage is accurate to ± 3 ft of pressure head, elevation data for the system may be of widely varying quality. The effort necessary to determine elevations in the system to an accuracy of ± 5 ft is worthwhile. Therefore static head in the system can be known to an accuracy of ± 8 ft (3.5 psi). With poor-quality data, adjustments to the model during calibration to achieve such accuracy may not be worthwhile. A better way of looking at calibration accuracy is in terms of head loss. If, for example, there is only 20 ft of head loss from the source to the perimeter of the system (typical of small towns in flat terrain), the accuracy of the gage and elevations will be the limiting factor, and an overall accuracy of ± 5 ft should be achieved. If the system has several hundred feet of head loss from its highest point to its perimeter, then an accuracy of ± 15 ft will be good.

Water use, tank water level, and pump operations are continually changing in a water distribution system. Thus, in order to obtain data to calibrate a model, it is necessary to obtain a "snapshot" of the flows and pressures in the system at a single point in time. Fortunately, tank water levels, and hence the overall level of the piezometric surface, change fairly slowly over time. Therefore, if pumps are not switched on and off and use does not significantly change, it is possible to collect pressure and flow data over a several hour period and still obtain a reasonably clear "snapshot" to the system.

Once the data are assembled, the question still remains as to what parameter should be adjusted to achieve calibration. The engineer is in the position of an individual attempting to adjust the color on a television set with several knobs. Because of the time and cost involved with running a model, the engineer cannot simply start adjusting parameters on a hit-or-miss basis. The question is: what parameter should be adjusted? Some engineers argue that the water use estimate should be adjusted because it is the number that is least certain. Others maintain that C is the weakest piece of information. Still others claim the boundary conditions should be adjusted. The answer is, obviously, that the parameter that is incorrect is the one that should be adjusted. How to identify that parameter, and the size of the adjustment, are discussed in a subsequent section.

Some engineers maintain that values of C found in the literature are sufficient for model use. This is generally not true since: (1) literature values are merely average values and there is a good deal of variation about the average; (2) in a model of a skeletal system, a pipe must also account for the carrying capacity of nearby mains which have been eliminated from the model; (3) the C factor for a pipe must also account for minor losses which are not explicitly accounted for using equivalent pipe lengths; and (4) water use in a model is lumped at nodes, while in the system use is actually spread between the nodes. In most cases adjustments to C are minor, but occasionally some surprising insights are gained about the system during model calibration. While it is the C factor values determined in calibration that will be used with the model when it is applied in predicting future pressures and flows, the water use distribution must also be determined as part of the calibration procedure. If the water use distribution is in error, the C factors will not be accurate.

Those who still doubt the need to precisely calibrate a model should consider the illustrative example of a one-pipe system given below.

EXAMPLE. Suppose there is a system with a single pipe and tank with the following characteristics:

Length = 2 miles = 10560 ft
Diameter = 10 in.
Flow = 537 gpm
Water level in tank = 325 ft
$C = 50$
Head at end of line = 270 ft.

Now suppose an engineer knows only the length, diameter and final head, and needs to predict final head when the flow is increased by 200 gpm and the tank is almost empty (level = 305 ft). The engineer could assume the tank is half full (level = 315 ft) and that the C factor is 80, which is typical of 40-year-old unlined cast iron mains. From these assumptions, the engineer could calculate that the flow in the pipe, using the Hazen-Williams equation, is 503 gpm:

$$Q = (80)(10)^{2.63} \left[(315 - 270)/(10.5)(10560) \right]^{0.54}$$

$$= 503 \text{ gpm.}$$

The engineer can now predict that when the flow is increased by 200 gpm the head

$$H = 305 - \frac{10.5(10560)}{10^{4.87}} \left(\frac{703}{80} \right)^{1.85}$$

$$= 222 \text{ ft.}$$

In reality the head at the end of the pipe would be

$$H = 305 - \frac{10.5(10560)}{10^{4.87}} \left(\frac{737}{50} \right)^{1.85}$$

$$= 184 \text{ ft.}$$

The difference between the correct answer and the one based on assumptions is 38 ft (17 psi), the difference between adequate and inadequate pressure in many cases. The engineer's error was caused by assuming the tank elevation and not making pressure readings over a wide enough range of flow to understand that the C factor was incorrect.

In the following sections, data collection for model calibration is described, a formula is derived for determining the adjustment to C and water use, and the implications of the formula and its shortcomings are discussed. The key to the approach presented in the following sections is that pressure readings should be taken at different flow rates while the same boundary conditions are held constant. This can be done by opening hydrants in the area of the test and measuring the resulting flow and pressure.

4.6.   DATA COLLECTION FOR CALIBRATION

With an unlimited budget, it would be possible for an engineer to measure the C factor for virtually every pipe in a distribution system and instantaneously measure water use throughout the system. While the latter measurement is becoming possible with telemetered water meter reading, the cost of measuring head loss and flow in pipes remains substantial.

The key to collecting data for model calibration, therefore, lies in getting enough data to give a "snapshot" of flows and pressures in the system. In order to decide whether to adjust the use estimate or the C factor, it is necessary to perturb the flows significantly by a known amount, while keeping the boundary conditions constant. This can be done by performing a fire flow test in which (1) the pressure at a hydrant is measured, (2) a nearby hydrant is opened and the flow is recorded, and (3) the pressure at the first hydrant is recorded while the second hydrant is open. The additional information required beyond the usual data collected during a fire flow test are: (1) the elevation of the pressure gage at the hydrant, and (2) the boundary conditions at the time of the test (i.e., water levels in nearby tanks and pressures at pumps). With this information plus the results of a few C-factor tests it is possible to do an accurate job of model calibration at reasonable cost.

Often the weakest pieces of calibration data are the elevations of key points in the system (e.g., pumps and hydrants where tests are conducted). Obtaining elevations from maps with 20 ft contour intervals produces data that is good to only ±10 ft. Elevations should be determined from maps or aerial photographs with finer contour intervals. Engineering drawings made at the time the system was constructed, or the local gas or sewer lines were laid, can serve as a better source of elevation data. As a last resort, it is usually not prohibitively expensive to survey the elevations of test hydrants, water tanks, and pumps from known benchmarks.

In addition to the fire flow and C-factor tests it is also very helpful to read flow meters on major mains at about the time the hydrant tests are conducted. This provides an excellent cross check on the flows. Similarly, the pressure at pressure reducing valves should always be checked. A valve may have been set to give a downstream pressure of 60 psi in 1965, but that is no reason to assume it is still delivering that pressure today. Spot checks of pressure throughout the system should be made to help locate any surprises (e.g., open connections with other utilities or other pressure zones or valves that have been accidentally left closed since a pipe break was repaired several years earlier).

The fire flow test should significantly lower pressure in the part of the system being studied. If it does not, it may be necessary to open two or more hydrants to achieve a drop in pressure of at least 5 but preferably 20 psi.

As mentioned earlier, data collection for systems with recording pressure gages or telemetered pressure data is quite easy. Nowadays the price of small recording pressure gages is fairly reasonable. The trick to installing them is to place them on a main line, not on a service line where pressure can fluctuate as a result of relatively small use changes, and to place them in vandalproof locations. This may require tapping a line inside a vault or manhole.

If a system is too large to study at one time (i.e., by the time the engineer drives from one test point to another the boundary conditions have changed), it is best first to divide the system into the pressure zones and, if these are still too big, divide the system into pie-shaped sectors and collect data for one sector at a time. The locations at which the fire flow tests are conducted should be at the perimeter of the system and near major water users. Finally, if the system can be supplied by either pumps or tanks, it is advantageous from a model calibration standpoint to gather calibration data when the tanks are the source. This eliminates potential problems such as the pump operating point shifting dramatically during a hydrant test.

Techniques for conducting hydrant tests and C-factor tests are described in greater detail in Chapter 8.

4.7. DEVELOPMENT OF ADJUSTMENT FACTORS FOR CALIBRATION

Knowing the results of hydrant flow test and the water level in a tank or pressure provided by a nearby pump, it is possible to derive a formula to determine the amount by which C or water use should be adjusted. The key to developing these correction factors is to be able to replace a portion of the distribution system with an equivalent pipe with a C and water use. This is why it was recommended that this procedure be applied to arteries in the system serving roughly pie-shaped sectors. This sector should have at one end a constant head node (i.e., a tank or a pump) and at the other end, the node corresponding to the hydrant flow test.

The head loss between the constant head node and the test hydrant node can be given by the Hazen–Williams equation for what will be called the low-flow (hydrant closed) and high-flow (hydrant(s) open) cases as

$$h_1 = K_1(S/C)^n \tag{4.7.1}$$

$$h_2 = K_2[(S + F)/C]^n \tag{4.7.2}$$

where
h_1 = head loss at low flow, L
h_2 = head loss at high flow, L
F = difference between flow at high and low flow, L^3/T
    (i.e., flow from hydrant(s))
C = C for equivalent pipe
K_1 = constant for head loss equation at low flow
K_2 = constant for head loss equation at high flow
n = exponent for head loss equation
S = sum of the water use at nodes served by the equivalent pipe, L^3/T,

$$= \sum_{i=1}^{m} Q_i$$

m = number of nodes served by equivalent pipe
Q_i = flow at ith node served by equivalent pipe, L^3/T.

There are four unknowns in the two equations above: K_1, K_2, C, and S. K_1 and K_2 depend on the lengths and diameters of the pipes that make up the equivalent pipe. They would be equal if there were no water use between the constant head node and the test hydrant, but since water users are distributed along the length of the complicated network being represented by the equivalent pipe, K_1 and K_2 will be different. The h values above refer to head loss between the nearby tank or pump and the test node.

K_1 and K_2 can be estimated fairly accurately utilizing the model user's initial estimates of C and S (referred to as C_e and S_e). Given the user's best estimates of C and S, the first run of the model can be used to generate estimates of head loss at low flow (h_3) and high flow (h_4). These estimates of h_1 and h_2 can be inserted into Eq. (4.7.1) and (4.7.2) to produce fairly accurate values of K_1 and K_2:

$$K_1 = h_3(C_e/S_e)^n \qquad (4.7.3)$$

$$K_2 = h_4[C_e/(S_e + F)]^n \qquad (4.7.4)$$

where

h_3 = model prediction of h_1 for low flow, L
h_4 = model prediction of h_2 for high flow, L
C_e = user's initial estimate of C
S_e = user's initial estimate of S, L^3/T.

Inserting the values of K_1 and K_2 into Eq. (4.7.1) and (4.7.2) and solving for S and C gives

$$S = \frac{F}{\dfrac{b}{a}\left(1 + \dfrac{F}{S_e}\right) - 1} = AS_e \qquad (4.7.5)$$

$$C = \frac{FC_e}{b(S_e + F) - aS_e} = BC_e \qquad (4.7.6)$$

where

$$a = \left(\frac{h_1}{h_3}\right)^{1/n}$$

$$b = \left(\frac{h_2}{h_4}\right)^{1/n}$$

$$A = \frac{F}{(b/a)(S_e + F) - S_e}$$

$$B = \frac{F}{b(S_e + F) - aS_e}.$$

The equations given above can be used to calculate improved values of C and water use. The parameters a and b are useful indicators of the magnitude of the error involved in using the initial estimates of C and S. The A and B parameters are the actual adjustments to water use and C. For example, an A of 1.15 means that the water use estimate should be increased by 15 percent over the initial estimate, while a B of 0.80 means that C should be reduced by 20 percent.

The correction factor should be applied to all pipes comprising the equivalent pipe and all nodes being served by that pipe. The engineer may want to use some judgment and correct some pipes or nodes by more than others based on prior knowledge (e.g., the C of one pipe may be known with relative certainty because of a recent test, or the flow at a node may be known fairly precisely from the results of a meter reading). In general, though, the correction factors should be used exactly as calculated above. Once the corrections have been made, the model should be rerun to make certain the accuracy is acceptable. It may be necessary to apply the correction factors several times to arrive at a satisfactory calibration.

EXAMPLE. Consider the skeletal network shown in Fig. 4.2, with the exact diameters, lengths, C's, and water uses given in the figure. Now suppose an engineer incorrectly estimates C to be 115 for all pipes and incorrectly estimates water use at nodes 20 and 30 as 150 and 400 gpm, respectively. The engineer knows the elevation of water in the tank to be 200 ft. The actual heads are shown in row 1 of Table 4.1, while the results of the model run using the engineer's initial estimates are shown in row 2. Water use and C values are summarized in Tables 4.2 and 4.3. Hydrant flow rates corresponding to the high flows are 1200 gpm at node 70 and 2500 gpm at node 40.

The first step is to determine the error by calculating a and b at both test nodes. The Hazen–Williams exponent 1.85 is used for n to give

$$a(40) = \left(\frac{200 - 181}{200 - 189}\right)^{0.54} = 1.34$$

$$a(70) = \left(\frac{200 - 173}{200 - 184}\right)^{0.54} = 1.33$$

$$b(40) = \left(\frac{200 - 150}{200 - 162}\right)^{0.54} = 1.16$$

$$b(70) = \left(\frac{200 - 64}{200 - 123}\right)^{0.54} = 1.36.$$

For this network, the results of the hydrant test at node 70 will be used to adjust the use at node 60 and 70 and C in pipes h and i. The results of the test at node 40 will be used to adjust the other nodes and pipes, except pipe g, which did not significantly affect either test.

For node 40,

$$A = \frac{2500}{(1.16/1.34)(2550 + 2500) - 2550} = 1.37$$

since $S = 150 + 400 + 500 + 1500 = 2550$ gpm, and

$$B = \frac{2500}{1.16(2550 + 2500) - 1.34(2550)} = 1.02.$$

Since B is approximately one, the C factors do not require adjustment but the water use at nodes 20, 30, 40, and 50 should be increased by 37 percent.

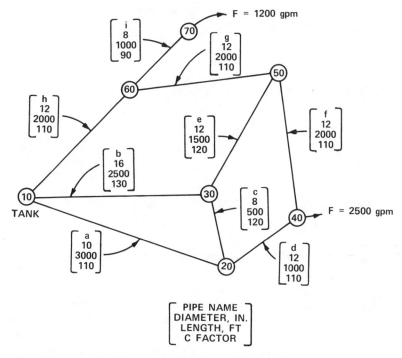

Fig. 4.2. Skeletal network for example.

Next, for node 70, A and B are found to be 0.95 and 0.72, respectively.

Since A is near one, there is no need to adjust water use, but the C factors must be decreased by 28 percent. These values are shown in the tables. When the corrected values are entered into the model the agreement between observed and predicted heads improves considerably. The new values for a and b are very close to one, indicating good calibration:

$$a(40) = 0.97, \quad a(70) = 0.96$$

$$b(40) = 0.97, \quad b(70) = 1.00.$$

Table 4.1. Hydraulic Grade Line (HGL) Data for Network Example (ft).

	Node 40		Node 70	
	Low Flow HGL	High Flow (2500 gpm) HGL	Low Flow HGL	High Flow (1200 gpm) HGL
Actual readings	181	150	173	64
Initial run	189	162	184	123
Corrected run	180	147	171	63

Table 4.2. *C* Factors for Network Example.

Pipe	Actual	Initial	Corrected
a	100	115	115
b	130	115	115
c	120	115	115
d	110	115	115
e	120	115	115
f	110	115	115
g	110	115	115
h	110	115	83
i	90	115	83

4.8. IMPLICATIONS OF CALIBRATION TECHNIQUE

The parameters a and b defined in the previous section are useful dimensionless indicators of the error in head loss at low and high flow, respectively. They convey a good deal more information than simply the difference between observed and predicted head loss. A value of a of 1.2 means the same thing for a large or small system, while an absolute difference between observed and predicted head of 10 ft may be quite good in a large system but poor in a very small system. For the purpose of this discussion a large system is one in which the head varies significantly (say more than 40 ft) between the high and low heads in the system. A small system is one with very little change in head across the system.

Values of a and b between 0.8 and 1.2 are achievable in virtually all cases, while values between 0.9 and 1.1 are possible in most. Values of a and b less than one indicate that the model overestimated head loss, while values greater than one indicate that it underestimated head loss. In general, a and b will vary from 0.5 to 2.0. Values outside that range indicate that the model has major problems related to boundary conditions or skeletonizing. The interpretation of a and b is summarized in Table 4.4.

Table 4.3. Water Use for Network Example (gpm).

Node	Actual	Initial	Corrected
20	500	150	205
30	2000	400	548
40	500	500	685
50	1500	1500	2055
60	1000	1000	1000
70	400	400	400
Total	5900	3950	4893

Table 4.4.   Interpretation of *a* and *b*.

Value	*a*	*b*
<1	Too much head loss predicted at low flow	Too much head loss predicted at high flow
$=1$	Head loss correct at low flow	Head loss correct at high flow
>1	Too little head loss predicted at low flow	Too little head loss predicted at high flow

While *a* and *b* provide insight into the nature of the error, the equations for
A and *B* must be used to determine the actual correction factors for *C* and
water use. To give the reader an appreciation of the effect of *a* and *b* on the
correction factor for *C* and water use (called *Q* in this section), the percentage
differences between initial and actual *C* and *Q* were calculated for a large array
of values of *a* and *b*, letting $C_e = 120$, $F = Q_e = 1000$ gpm be initial estimates,
for a simple, one-node system.

Figure 4.3 shows the effect of *a* and *b* on *C*. Within the area titled "Accept
C," *C* is within 10 percent of C_e. The dashed line corresponds to combinations

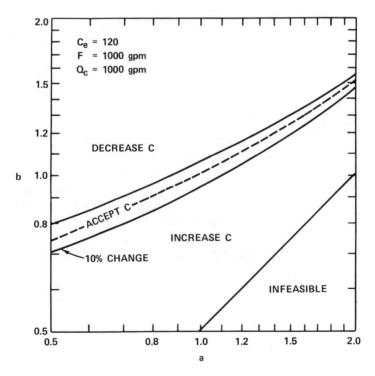

Fig. 4.3. Effect of *a* and *b* on correction to *C*.

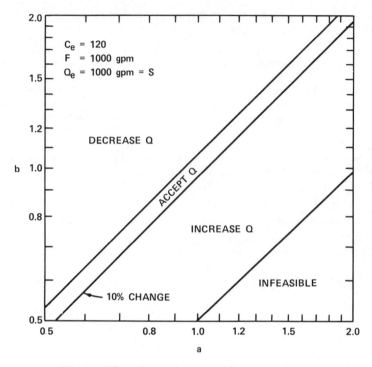

Fig. 4.4. Effect of a and b on correction to water use.

of a and b for which $C = C_e$, (i.e., no error in initial estimate). Along that line, a and b are related by $2b = a + 1$, which can be determined by letting $C = C_e$ in Eq. (4.7.6). The region titled "Infeasible" corresponds to combinations of a and b for which the denominator in the corresponding equation is negative.

Figure 4.4 shows the effect of a and b on Q. Within the area titled "Accept Q," Q is within 10 percent of Q_e, and the line corresponding to no change in Q is represented by $a = b$. For $b > a$ the water use estimate should be reduced, while for $b < a$ the water use estimate should be increased.

Figure 4.5, developed by overlaying Fig. 4.3 and 4.4, summarizes the nature of adjustments to Q_e and C_e to achieve calibration. Those who believe in calibration by adjusting C only are working under the assumption that $a = b$, while those who believe only Q should be adjusted have implicitly assumed that

$$b = \frac{aQ_e + F}{Q_e + F}. \tag{4.8.1}$$

Neither assumption is valid in general.

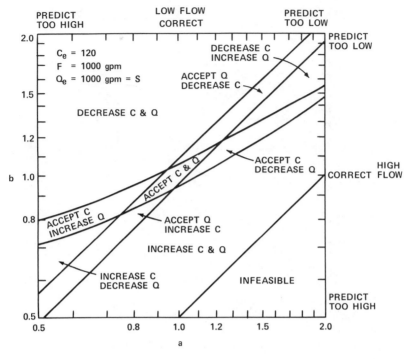

Fig. 4.5. Relation between type of error (*a* or *b*) and type of correction (*C* or *Q*).

There are a few special cases which provide insight into the type of adjustments needed for calibration:

1. If $a = b$ (similar percentage error at high and low flow), adjust C only.
2. If $a = 1$ and $b \neq 1$ (good calibration at low flow only), adjust both Q and C by roughly similar fractions.
3. If $b = 1$, $a \neq 1$, and $F > Q_e$ (good calibration at high flow; low flow much less than fire flow), adjust Q only.

These statements can be proved by setting a and/or b to the appropriate values and reducing Eqs. (4.7.4) and (4.7.5). This is left to the reader as an exercise.

The importance of insuring that F is significant in comparison to Q can be shown by plotting the line for $C = C_e$ for several values of F/Q_e, as shown in Fig. 4.6. As F approaches 0, the line corresponding to $C = C_e$ approaches the line for $Q = Q_e$, and it is no longer possible to decide if C or Q should be adjusted. In practical terms, as F approaches Q_e, the calibration becomes

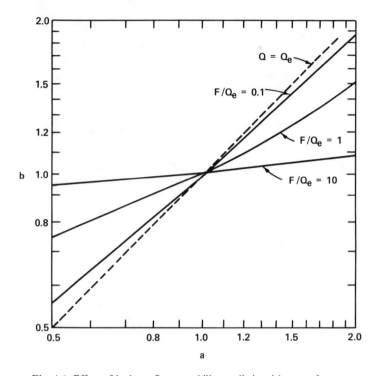

Fig. 4.6. Effect of hydrant flow on ability to distinguish type of error.

increasingly sensitive to errors in measuring h_1 and h_2. The accuracy of the adjustment factors is greatest for large flows through the hydrants.

Ideally, the initial estimates of C and Q used in the first run of the model are so accurate that there is no need to correct the input data using the equations given above. In most cases, however, some adjustment is necessary, but usually the corrections are small. When the required adjustments are large, the engineer should not blindly change C or Q without trying to understand why the initial estimates were poor. Taking time to understand the source of errors can provide valuable insights into the behavior of water distribution systems.

If the corrected C is much lower that the initial estimate, it is implied that the hydraulic carrying capacity of the mains is less than anticipated. While this may be due to pipe roughness increasing unusually rapidly with pipe age, it may also be due to a closed or partially closed valve on a main. It is not uncommon in a modeling study to locate valves that have been mistakenly left closed.

A corrected C that is much greater than the initial estimate implies that the hydraulic carrying capacity of the mains is greater than anticipated. This is often the result of not including fairly important mains in developing the skeletal model. This can be corrected by increasing C or including some of the pipes that had been eliminated.

If the corrected water use is much less than the initial estimate, the engineer should attempt to identify large water users not operating when the data were collected. For example, field observations may have been made on a school holiday, or workers at a nearby water-using industry may have been on strike. To achieve calibration, the water uses should be adjusted to reflect actual use on the day the data were obtained, but these values should not necessarily be used as a basis for flows when the model is used in a predictive manner.

If the corrected water use is much greater than the initial estimate, the engineer should look for unusual water users that the model may be overlooking. For example, the municipal swimming pool may have been filled that day, or lawn watering may have been especially intense due to a drought. Large unexplained uses may also be the result of leakage or illegal connections.

In no case should unrealistic adjustments to values for C or water use be made simply to achieve calibration. If calibration can be achieved only with ridiculous values for input data, the field observations, especially the boundary conditions, should be checked for accuracy.

The procedure for calibrating models described in Section 4.7 is not fool-proof. There are problems in systems in which the head loss is very small during normal use periods, so that both h_1 and h_3 are virtually zero at normal use. In this situation, a becomes essentially $0/0$ and tiny errors in measurements result in ridiculous values for C and water use. For these systems, the model must be calibrated using only data collected during fire flow tests as data collected during normal use periods will make the model appear to be calibrated for virtually any combination of C and water use.

If it becomes necessary to use a trial-and-error approach, it is important to record the results in a master table, which should include the setting of each calibration parameter and predicted head at key points for every run. This will make comparisons with measured head and final selection of values much easier.

A summary of the steps involved in calibrating a water distribution system model using the method described in section 4.7 is shown in Fig 4.7.

4.9. PRODUCTION RUNS OF THE MODEL

Once the model is calibrated for high and low flows, and possibly even several boundary conditions, the engineer can use the model to analyze the behavior of the system for different conditions and predict how the system will behave with various improvements. The engineer should set up several data files with use estimates corresponding to different planning scenarios and time

OVERVIEW

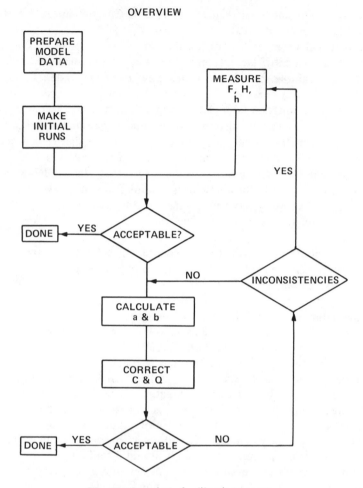

Fig. 4.7. Overview of calibration process.

windows. These files should be merged with the network file for the production runs.

Before plunging into production runs, the user should outline the runs to be made and establish a numbering or naming scheme for each run. It is much easier to talk about "Run 17" than to recite "the run with the 24-in. pipe between First St. and Pine Rd., with the new 1000 gpm pumps at Clear Creek and the 100 ft tank on Brown's Hill for 2010 population summer evening use with no conservation measures."

A good way of keeping track of the runs is a master table with the run number, time window, projection scenario, construction phases, etc. identified in the columns of the table for each run. This will help to insure that a reasonably large, significantly different array of alternatives has been examined.

Output from the various runs should be labeled and kept together in a binder. Copious notes should be made on the printouts concerning such things as adequacy of the new pipe simulated and whether this is an improvement over previous runs. Where new construction is being investigated, all items to be constructed for each alternative should be identified. This will make it easy to make planning level cost comparisons later.

This description of the runs should be in sufficient detail that the runs can be examined several weeks later and still be understandable. It is amazing how quickly after the runs are made they all begin to look the same. Yet after a study is complete, the engineer can almost always count on being asked such questions as, "What if we use an 18-in. pipe on Union St. instead?" For this reason, the program data and a copy of the version of the model used for the study should be stored on tape indefinitely, along with instructions on the format of the tape.

When a small number of candidate designs have been selected, they should be screened not only for a single water use rate (usually worst case—high flow), but for the entire range of operating conditions: low flow to be sure the tanks can fill and to find areas with excessive pressure; average flow to see if the net flow in the system is along the desired paths and fire flows at a few critical locations to see if pressures will be adequate. In general, the size of small neighborhood lines will be controlled by fire flow requirements, while the size of major trunk mains will be controlled by instantaneous peak flow. Tank sizing can be checked easily with extended period simulation models.

REVIEW QUESTIONS

1. What problems can occur when a water distribution system is calibrated for only average flow conditions?
2. Why is it necessary to know the boundary conditions when using the results of hydrant flow test to calibrate a model?
3. Consider a water distribution system with which you are familiar. Where would you want to conduct hydrant flow test for model calibration?
4. Name five conditions you should simulate in using a model to assist in designing a water distribution system.
5. Which type of network model is best for a branched system with one source?
6. Name a situation in which an extended period simulation should be used to study a design.
7. Why should the water use estimate used for calibration not be the same as that used to design for future conditions?
8. Consider how you would write a computer program that is capable of identifying loops given only the beginning and ending node for each pipe. Consider the work involved in doing this by hand.

9. Why are pressure reducing valves difficult to consider in a model?
10. Why is it somewhat misleading to discuss calibration accuracy only in terms of absolute differences between observed and predicted heads?
11. Why should hydrant discharges be fairly large during tests for model calibration?
12. Is it better to adjust water use or C estimates to achieve calibration?

PROBLEMS

1. Prove the following statements that were made in Section 4.8:
   a. If $a = b$, adjust C only;
   b. If $a = 1$ and $b \neq 1$, adjust both C and Q similarly;
   c. If $b = 1$ and $a \neq 1$, adjust Q only.
2. The pressure in a system is 63.4 psi at normal flow and 43.2 psi when a hydrant is discharging 1200 gpm. The head at the source is 200 ft above the elevation of the hydrant. In calibrating a distribution systems model, you estimated water use at 5 mgd and $C = 100$. The model predicted pressures of 66.5 and 52.8 psi for normal flow and fire flow of 1200 gpm, respectively. Correct C and water use to achieve calibration.
3. Knowing that the head loss between a water source and a test hydrant connected by 1000 ft of 8-in. pipe is 10 ft under normal use, plot the possible combinations of flow and C that can result in this loss. Plot use on the vertical axis and C on the horizontal axis.

   Now suppose the head loss increases to 40 ft when the hydrant discharges 800 gpm. Plot the combinations of C and Q that satisfy this condition on the same graph.

   What pair of C and Q values is correct for both cases?

REFERENCES

AWWA, 1975, Sizing Water Service Lines and Meters, *AWWA Manual M22*, Denver, CO.

Boland, J.J., D.D. Baumann, and B. Dziegielewski, 1981, An Assessment of Municipal and Industrial Water Use Forecasting Techniques, IWR CR-81-C05, U.S. Army Institute for Water Resources, Ft. Belvoir, VA.

Clark, J.W., W. Viessman and M.J. Hammer, 1977, *Water Supply and Pollution Control*, IEP–Dunn-Donnelley Publishing, New York.

Collins, M.A., 1980, "Pitfalls in Pipeline Network Analysis Techniques," *Transportation Engineering J. ASCE*, Vol. 106, No. TE5, p. 507.

Eggener, C.L., and L. Polkowski, 1976, "Network Models and the Impact of Modeling Assumptions," *J. AWWA*, Vol. 68, No. 4, p. 189, April.

Linaweaver, F.P., J.C. Geyer, and J.B. Wolff, 1967, "Summary Report on Residential Water Use Research Project," *J. AWWA*, Vol. 59, No. 3, p. 269, March. (See complete series of reports on Residential Water Use available from The Johns Hopkins Univ.).

Williams, G.S., and A. Hazen, 1920, *Hydraulic Tables*, John Wiley & Sons, New York.

5 | SIZING WATER MAINS

5.1. INTRODUCTION

In every problem involving selection of pipe sizes, there is some size or combination of sizes which will result in the lowest life-cycle cost for the pipe(s), associated pumping equipment, and energy. The most precise way of determining the least-cost pipe size is to perform detailed design and cost estimates for an exhaustive array of possible alternatives. This is prohibitive from a time and cost standpoint, especially in the early stages of a study when the number of alternatives is large.

Since the savings that can be realized by selecting the least-cost pipe size are significant, there is a need for a simpler method that will still yield correct answers. Possible approaches can be grouped into several categories: (1) feasible solution, (2) trial-and-error optimization, (3) rules-of-thumb, and (4) mathematical optimization. In the feasible solution approach, one searches for a set of pipe sizes that will function hydraulically without concern for the least-cost solution. In trial-and-error optimization, one compares alternative feasible pipe sizes to determine the least cost by trial and error. In using rules of thumb, one selects pipe sizes on a criterion [e.g., 1.8 m/s (6 ft/s) at peak flow] and checks to insure that the solution is feasible hydraulically. In mathematical optimization, one uses a mathematical programming technique (e.g., linear programming, integer programming) to determine an optimal solution.

The first three approaches can be performed manually but do not guarantee that a least-cost solution will be found. Mathematical optimization can yield a least-cost solution but requires a computer and some expertise in mathematical programming, and there are difficulties in applying the methods to large systems. This chapter focuses on providing fairly simple, manual pipe sizing procedures that can be used to select least-cost pipe sizes for either series or branched systems. Typical applications are classified in the following section; then methods are developed for pumped systems and gravity systems.

Some rules of thumb for pumped systems are presented. Finally, the use of mathematical programming techniques to achieve optimal solutions for looped systems is described.

5.2. CLASSIFICATION OF PROBLEMS

Before developing a method it is useful to classify problems according to level of difficulty. Such a hierarchy is shown in Fig. 5.1. The most important distinction between different types of pipe sizing is whether the pipes are part of a looped system. If the pipe is to be part of a highly looped system, the pipe size will determine the flow the pipe can carry. However, since the flow to be carried is not known beforehand, it is difficult to optimize pipe size without using a very complicated optimization technique. In this type of situation, most designers rely on a trial-and-error approach. In some looped systems with fairly large loops, it is possible to identify a design flow and use the simpler techniques available for non-looped systems.

Within non-looped systems, sizing problems can be classified according to whether there is: (1) a single flow rate along the pipe, (2) a decreasing flow rate along a principal pipe as smaller lines split off, or (3) a branched network in which it is not possible to identify a principal pipe. In the first case (single pipe), if the pipe is fed from a gravity source, it is a simple hydraulic problem to select the minimum diameter pipe that will meet the need. If the pipe is fed from a pump, the problem involves a tradeoff between pipe cost and pump equipment and energy cost. Techniques to identify optimal pipe size have been

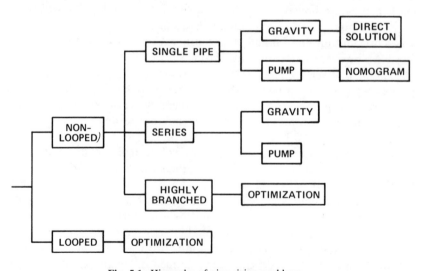

Fig. 5.1. Hierarchy of pipe sizing problems.

developed by Dances (1977), Deb (1973), Osborne and James (1973), Stephenson (1976), and Walski (1982).

When the flow changes along a pipeline because water is withdrawn along the way, there is a point at which the flow decreases to the extent that it is economical to switch to a smaller diameter pipe. The problem becomes one of sizing pipes in series. Some of the better techniques for solving this kind of problem have been developed by Cowan (1971), Deb (1976), and Swammee, Kumar, and Khanna (1973). Sections 5.4 and 5.5 deal with modification of the above techniques to produce a simple pipe sizing method.

If a system is branched to the extent that it is cumbersome to consider it as a few major lines with minor branches, it becomes necessary to use more sophisticated mathematical programming models. The best of these techniques are those presented by Schaake and Lai (1969), Watanatada (1969), Shamir (1974), Rasmusen (1976), Alperovits and Shamir (1977), Gessler (1982), Morgan and Goulter (1982), Quindry, Brill and Liebman (1981), Featherstone and El-Jumaily (1983), and Rowell and Barnes (1982). Most of these are applicable to looped systems as well. All require a computerized solution. Shamir (1979) has summarized the techniques.

Overall, looped or highly branched systems must be addressed using a sophisticated optimization technique or a trial-and-error approach. In systems that can be considered as pipes in series, there are ways of arriving at an optimal solution with manual calculations. These techniques are different for pumped and gravity systems. They are presented in the following sections.

5.3. COSTS

The costs involved in constructing and operating a water pipe with associated tanks and pumps include:

1. Pipe
2. Pumping equipment
   a. Structural
   b. Mechanical
3. Energy
   a. Lift
   b. Friction
4. O&M labor and supplies
5. Storage tanks.

Of the above cost components only pipe, and energy to overcome friction losses are significantly affected by the selection of pipe diameter. The effect on pumping station construction costs only becomes significant when the selection of pipe diameter alters the number of pumping stations required. If the number of stations does not change, then the change of pumping equipment cost due to a change in pipe diameters is insignificant compared with the change in pipe and energy costs.

The relationship between pipe diameter and pipe cost can be described by equations of the form

$$c_c = AD^b \qquad (5.3.1)$$

or

$$c_c = A \ \exp \ (bD) \qquad (5.3.2)$$

where

c_c = unit capital cost of pipe installed, $/ft ($/m)
D = pipe diameter, in. (mm)
a,b = regression coefficients
A = cost coefficient corrected for price level, $a(ENR)/3800$
ENR = appropriate value of ENR construction cost index.

The regression coefficients a and b depend on the type of pipe, depth of cover, amount of rock excavation, and pipe pressure rating among other factors. Values were determined for typical conditions in mid-1982 dollars (ENR = 3800) and are shown in Fig. 5.2 and Table 5.1. For small pipes $D < 10$ in. (250 mm) costs are based on PVC pipe and are best represented by equation (5.3.2). For mid-sized pipe 8 in. (200 mm) $\leqslant D \leqslant 48$ in. (1200 mm), the costs are based on ductile iron pipe and are represented by Eq. (5.3.1). For large pipe, $D \geqslant 48$ in. (1200 mm), costs are based on prestressed concrete cylinder pipe and are also represented by Eq. (5.3.1).

The costs given in Table 5.1 are merely for typical situations and should be replaced by costs for the particular study area if the engineer has such data. Readers are cautioned against using the cost functions given above for atypical situations. At a minimum, the cost functions should be verified using actual data from the study area. The results of any analysis are only as good as the cost data.

Annual energy costs per unit of pipe length are based on average flow and can be given by

$$c_e = K_1 P(rQ)h/e \qquad (5.3.3)$$

where

c_e = cost of energy, $/yr/ft ($/yr/m)
P = price of energy, $/kwhr

Q = peak flow rate, MGD or gpm (m³/s)
h = friction head to be overcome, ft/ft (m/m)
r = average to peak flow ratio
e = wire-to-water efficiency expressed as decimal

$$K_1 = \begin{cases} 11.4 \text{ for } Q \text{ in MGD, } c_e \text{ in \$/yr/ft} \\ 0.0164 \text{ for } Q \text{ in gal/min, } c_e \text{ in \$/yr/ft} \\ 860 \text{ for } Q \text{ in m}^3/\text{s, } c_e \text{ in \$/yr/m.} \end{cases}$$

The head loss can be given by an equation of the form

$$h = K_2 \frac{(rQ)^{m-3}}{D^m} \tag{5.3.4}$$

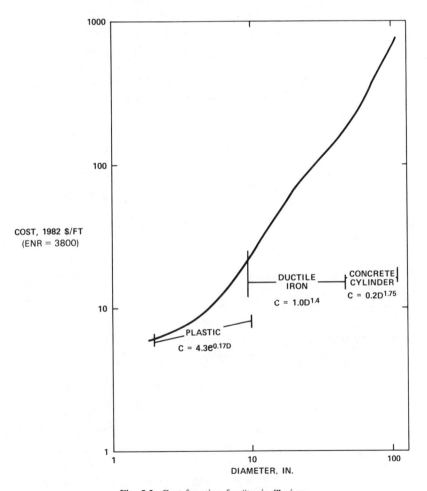

Fig. 5.2. Cost function for "typical" pipe.

Table 5.1. Coefficients for Pipe Cost Function.

	$a*$, \$/ft, D in in.	b	Equation Type
Small pipe	4.3	0.17/in.	(5.3.2)
Mid-size pipe	1.0	1.4	(5.3.1)
Large pipe	0.20	1.75	(5.3.1)

*Mid-1982 dollars.

where

$$D = \text{internal diameter, in. (mm)}$$

$$m = \begin{cases} 5, & \text{Darcy–Weisbach Equation} \\ 4.8, & \text{Hazen–Williams Equation} \end{cases}$$

$$K_2 = \begin{cases} f\,(5652), & D \text{ in in., } Q \text{ in MGD} \\ 10.4/C_w^{1.85}, & D \text{ in in., } Q \text{ in gal/min} \\ 1.88 \times 10^6/C_w^{1.85}, & D \text{ in in., } Q \text{ in MGD} \\ f\,(3.144 \times 10^{13}), & D \text{ in mm, } Q \text{ in m}^3/\text{s} \end{cases}$$

$$f = \text{friction factor (Darcy–Weisbach)}$$

$$C_w = \text{Hazen–Williams } C \text{ factor.}$$

Combining Eqs. (5.3.3) and (5.3.4) gives

$$c_e = \frac{K_3 P (rQ)^{m-2}}{D^m e} \tag{5.3.5}$$

where

$$K_3 = K_1 K_2.$$

5.4. PUMPED SYSTEMS

In pumped systems, the minimum cost can be determined as a tradeoff between pipe cost and energy costs. The total present worth cost function can be developed for larger [Eq. (5.3.1)] pipe sizes by minimizing the sum of these costs for each length of pipe (since other costs are relatively independent of diameter):

$$T = AD^b + ED^{-m} \tag{5.4.1}$$

where

$$T = \text{total cost, \$/m}$$

$$E = SK_3 P \,(rQ)^{m-2}/e$$

$$S = \text{series present worth factor, years.}$$

Taking the first derivative with respect to diameter, and setting the result to zero yields

$$\frac{dT}{dD} = 0 = AbD^{b-1} - EmD^{-(m+1)}.$$ (5.4.2)

Solving for D yields

$$D = \left(\frac{Em}{Ab}\right)^{1/(b+m)}$$ (5.4.3)

which gives optimal pipe size.

The second derivative can be given by

$$\frac{d^2T}{dD^2} = Ab(b-1)D^{b-2} + Em(m+1)D^{-(m+2)}.$$ (5.4.4)

For A, E, and m greater than zero and b greater than one, dT^2/dD^2 is positive, indicating that the optimum pipe size is a minimum.

For small diameter pipes the capital cost of the pipe is given by Eq. (5.3.2) and the total cost function becomes

$$T = A \exp (bD) + ED^{-m}.$$ (5.4.5)

Differentiating yields

$$\frac{dT}{dD} = 0 = Ab \exp (bD) - EmD^{-(m+1)}.$$ (5.4.6)

Solving for D yields

$$D = \left(\frac{Ab \exp (bD)}{Em}\right)^{-1/(m+1)}$$ (5.4.7)

which must be solved iteratively for D. The second derivative

$$\frac{d^2T}{dD^2} = Ab^2 \exp (bD) + Em(m+1)D^{-(m+2)}$$ (5.4.8)

is positive as long as A, E, and m are positive.

Once D is found, the problem is still not solved, since pipe can only be purchased in certain commercially available nominal diameters. Converting continuous pipe diameter D into a nominal diameter d where the continuous diameter D lies between two nominal diameters d_j and d_{j+1} ($d_j < D < d_{j+1}$) involves finding the diameter for which the cost is lower. That is, select d_j if

$$Ad_j^b + Ed_j^{-m} < Ad_{j+1}^b + Ed_{j+1}^{-m} \tag{5.4.9}$$

and select d_{j+1} otherwise. Eq. (5.4.9) can be rearranged to give

$$Z = \left(\frac{d_j}{d_{j+1}}\right)^m \left[\frac{d_{j+1}^{b+m} + (E/A)}{d_j^{b+m} + (E/A)}\right] \tag{5.4.10}$$

and the criteria becomes: Select d_j if $Z > 1$ and d_{j+1} if $Z < 1$.

The procedure for selecting least-cost pipe sizes involving a trade-off is shown in Fig. 5.3 and described below:

1. Determine water use at each junction for peak and average flow, power cost (P), system efficiency (e), head loss coefficient (C_w or f) and exponent (m), price level (ENR), cost function coefficients (a, b), series present worth factor (S), and minimum head at junctions ($\overline{H}_i$).

2. Based on water uses, calculate flow for peak (Q) and average (rQ) conditions, and $E_i = SK_3 P(rQ)^{m-2}/e$ for each pipe and correct for price level [$A = (\text{ENR}/3800)a$].

PUMPED LINE

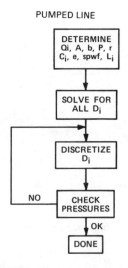

Fig. 5.3. Procedure for pumped system.

3. Calculate optimal continuous diameter (D_i) for each pipe using

$$D_i = \left(\frac{E_i m}{Ab} \right)^{1/(b+m)}$$ (5.4.11)

for medium and large pipes and

$$D_i = \left(\frac{E_i m}{Ab \ \exp \ (bD_i)} \right)^{1/(m+1)}$$ (5.4.12)

for small pipes (using iterative solution).

4. Round off diameter D_i, where $d_j < D_i < d_{j+1}$, by calculating

$$Z = \left(\frac{d_j^m}{d_{j+1}} \right)\left[\frac{d_{j+1}^{b+m} + (E/A)}{d_j^{b+m} + (E/A)} \right]$$ (5.4.7)

If $Z > 1$, pick d_j; if $Z < 1$, pick d_{j+1}.

5. Find head required at pump station(s), by determining the maximum value of H_{0I}

$$H_{0I} = \bar{H}_I + \sum_{i=1}^{I} \left(K_{2i} \frac{(rQ_i)^{m-3}}{d_i^m} \right), \qquad I = 1, 2, \ldots, n$$ (5.4.8)

where

$\bar{H}_I$ = head required at end of pipe I, ft (m)
H_{0I} = head required at pump to meet H_I at junction I, ft (m).

6. If the head at the pumps is excessive, round off marginal D_i's (upstream of largest $\bar{H}_I$) upward until max H_{0I} (pump pressure) is acceptable. If this is not possible, consider using two pump stations or installing storage at the end of the line to reduce peak flow, if peak and average flow differ greatly. If only one H_{0I} is excessive, it may be possible to lower the head requirement at that junction slightly.

EXAMPLE. Consider the system shown in Figure 5.4, and the data in Table 5.2. To find the optimal diameter it is first necessary to determine A and b for the cost function. Since pipe diameters greater than 1 ft are required, an a of 1.0 and b of 1.4 are appropriate. A can be determined as

$$A = 1.0 \left(\frac{4200}{3800} \right) = 1.10.$$

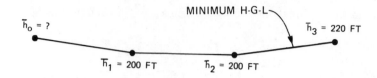

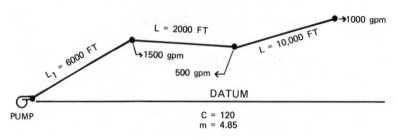

Fig. 5.4. Example of pumped line.

The peak flow in each pipe can be given as

$$Q_3 = 1000 \text{ gpm}, \qquad Q_2 = 1500 \text{ gpm}, \qquad Q_1 = 3000 \text{ gpm}.$$

E can be calculated for each pipe as

$$E_i = \frac{10.4(10)\ [(0.7)Q_i]^{2.85}\ (20)\ (0.0164)}{120^{1.85}\ (0.6)}$$

$$= 0.0029 Q_i^{2.85}.$$

$$E_1 = 23.8 \times 10^6, \qquad E_2 = 3.30 \times 10^6, \qquad E_3 = 1.04 \times 10^6.$$

The diameters can be determined by

$$D_i = \left[\frac{(4.85)E_i}{(1.10)\ (1.40)} \right]^{0.16} = 1.20 E_i^{0.16}$$

Table 5.2.   Data for Pumped System Example.

$P = 10¢/\text{kwhr}$
$e = 0.60$
$\text{ENR} = 4200$
$r = 0.70$
$S = 20$

which yields

$$D_1 = 18.2 \text{ in.}, \qquad D_2 = 13.2 \text{ in.}, \qquad D_3 = 11.0 \text{ in.}$$

These can be rounded to the commercial pipe sizes. Since 18.2 is between 18 and 20, $(E/A = 23.8 \times 10^6/1.1 = 21.6 \times 10^6)$,

$$Z_1 = \left(\frac{18}{20}\right)^{4.85}\left[\frac{20^{6.25} + 21.6 \times 10^6}{18^{6.25} + 21.6 \times 10^6}\right] = 1.02 \quad \text{use} \quad d_1 = 18 \text{ in.}$$

$$Z_2 (12, 14) = 0.97 \quad \text{use} \quad d_2 = 14 \text{ in.}$$

$$Z_3 (10, 12) = 0.98 \quad \text{use} \quad d_3 = 12 \text{ in.}$$

Next calculate the head loss in each pipe at peak flow

$$h_i = \frac{10.4 L_i}{d_i^{4.85}}\left(\frac{Q_i}{C_i}\right)^{1.85}$$

$$h_1 = \frac{10.4 (6000)}{18^{4.85}}\left(\frac{3000}{120}\right)^{1.85} = 19.6 \text{ ft}$$

$$h_2 = \frac{10.4 (2000)}{14^{4.85}}\left(\frac{1500}{120}\right)^{1.85} = 6.14 \text{ ft}$$

$$h_3 = \frac{10.4 (10,000)}{12^{4.85}}\left(\frac{1000}{120}\right)^{1.85} = 30.6 \text{ ft}$$

The pump head required to meet these flows is

$$H_{01} = 200 + 19.6 = 219.6 \text{ ft}$$

$$H_{02} = 200 + (19.6 + 6.14) = 225.7 \text{ ft}$$

$$H_{03} = 220 + (19.6 + 6.14 + 30.6) = 276.3 \text{ ft.}$$

Therefore the pumps must produce 276.3 ft at 3000 gpm. Since this head is reasonable the solution is optimal.

5.5. GRAVITY SYSTEMS

In a gravity flow system, the problem is one of determining the minimum pipe size such that, given the head at the source, the head at the end of the pipe is acceptable. The term *gravity system* refers to the fact that the head is fixed at the beginning of the line. The problem consists of selecting a pipe to convey the design flow with acceptable head loss. It does not refer to pipes not flowing full, as in sewer lines. The problem for large pipes is therefore

$$T = \sum_{i=1}^{n} AL_i D_i^b \qquad (5.5.1)$$

subject to

$$H_0 - H_n \geqslant \sum_{i=1}^{n} F_i D_i^{-m} \qquad (5.5.2)$$

where

$$n = \text{number of pipes in series}$$
$$H_0, H_n = \text{initial and final head, ft (m)}$$
$$F_i = K_2 L_i (Q_i)^{m-3} \text{ for the } i\text{th pipe}$$
$$L_i = \text{length of } i\text{th pipe, ft (m).}$$

The simplest way to solve the problem is to let the inequality constraint (5.5.2) be an equality constraint, since this will result in the lowest value of T (no wasted head). If Eq. (5.5.2) is an equality, it is possible to write the Lagrangian function

$$L = \sum_{i=1}^{n} AL_i D_i^b + y \left(\sum_{i=1}^{n} F_i D_i^{-m} - H_0 + H_n \right) \qquad (5.5.3)$$

where

$$L = \text{Lagrangian Function}$$
$$y = \text{Lagrangian multiplier.}$$

Differentiating Eq. (5.5.3) by each D_i and by y and setting the results to zero yields n equations of the form

$$\frac{dL}{dD_i} = 0 = AbL_i D_i^{b-1} - my F_i D_i^{-(m+1)}, \quad i = 1, 2, \ldots, n \qquad (5.5.4)$$

and one equation of the form

$$\frac{dL}{dy} = 0 = \sum_{i=1}^{n} F_i D_i^{-m} - (H_0 - H_n). \qquad (5.5.5)$$

This is a system of $n + 1$ equations and $n + 1$ unknowns (n D_i's and y). Solving Eq. (5.5.4) for values not a function of the pipe segment yields

$$\frac{my}{Ab} = \frac{L_i D_i^{b+m}}{F_i}.$$ (5.5.6)

Eq. (5.5.6) implies that if one of the D_i's (say D_1) is known, it is possible to determine the other D_i's from

$$D_i = \left(\frac{L_1 F_i}{L_i F_1}\right)^{1/(b+m)} D_1.$$ (5.5.7)

The value of D_1 can be determined by substituting for all of the D_i's in Eq. (5.5.5) and solving for D_1 as

$$D_1 = \left\{\frac{\sum_{i=1}^{n}\left[F_i\left(\frac{F_1 L_i}{F_i L_1}\right)^{m(b+m)}\right]}{H_0 - H_n}\right\}^{1/m}$$ (5.5.8)

Once D_1 is known, the other D_i values can be determined from Eq. (5.5.7). The value of the Lagrangian multiplier y can be determined from Eq. (5.5.6) as

$$y = \frac{Ab L_i D_i^{b+m}}{m F_i}$$ (5.5.9)

for any i. y is the rate of change of cost with respect to a change in $(H_n - H_0)$. As H_n approaches H_0 the total cost will increase. y will always be positive, since m, A, $b - 1$, and F_i are positive.

For the cases in which the small pipe cost function is more appropriate, the Lagrangian function in Eq. (5.5.3) becomes

$$L = \sum_{i=1}^{n} AL_i \exp(bD_i) + y\left(\sum_{i=1}^{n} F_i D_i^{-m} - (H_0 - H_n)\right).$$ (5.5.10)

Differentiating with respect to D_i and y yields

$$\frac{dL}{dD_i} = 0 = AbL_i \exp(bD_i) - myF_i D_i^{-(m+1)}$$ (5.5.11)

$$\frac{dL}{dy} = 0 = \sum_{i=1}^{n} F_i D_i^{-m} - (H_0 - H_n).$$ (5.5.12)

The relationship for generating the D_i's becomes

$$D_i = \left\{ \frac{F_i L_1}{F_1 L_i} \exp\left[b(D_1 - D_i)\right] \right\}^{1/(m+1)} D_1. \tag{5.5.13}$$

To determine D_1, substitute (5.5.13) for each D_i except D_1 in Eq. (5.5.12) and solve for D_1

$$D_1 = \left(\frac{\sum\limits_{i=1}^{n} F_i \left\{ \frac{F_i L_1}{F_1 L_i} \exp\left[b(D_1 - D_i)\right] \right\}^{-m/(m+1)}}{H_0 - H_n} \right)^{1/m} \tag{5.5.14}$$

Eqs. (5.5.13) and (5.5.14) must be solved iteratively for D_1. To minimize the number of iterations, it is helpful to determine a good set of initial values for the D_i's. This is done by assuming the head loss rate is constant in each pipe such that

$$\frac{F_i}{L_i D_i^m} = \frac{H_0 - H_n}{\sum\limits_{i=1}^{n} L_i} \tag{5.5.15}$$

which can be reduced to

$$D_i = \left[\frac{F_i \sum\limits_{i=1}^{n} L_i}{(H_0 - H_n) L_i} \right]^{1/m} \tag{5.5.16}$$

It is not necessary to iterate more than once or twice since the results are rounded to the nearest nominal pipe diameter.

In the case of gravity systems, it is not possible to optimize one pipe at a time as was done with the pumped system. The approach followed in this case is to round off all of the pipes except two (called i and I) to the nearest commercially available pipe size. When this is done, there is a certain amount of head available which can be lost in pipes i and I. This can be written

$$h^* = H_0 - H_n - \sum\limits_{\substack{k=1 \\ k \neq i,I}}^{n} F_k/d_k^m \tag{5.5.17}$$

where $h^* = $ head available for pipes i and I, m (ft). Therefore the head loss in pipe i and I must be less than $h*$:

$$h* > F_i d_i^m + F_I d_I^m. \tag{5.5.18}$$

The solution involves finding the least cost pair of pipe sizes (d_i, d_I) that satisfies the constraint in Eq. (5.5.18).

This solution can be shown graphically in Fig. 5.5. Eq. (5.5.18) divides the possible pipe sizes into a feasible and infeasible region. The equation of the dividing line is

$$D_I = \left(\frac{F_I}{h* - F_i/D_i^m} \right)^{1/m}. \tag{5.5.19}$$

The next step is to find the least-cost pipe size. This can be done graphically by plotting a cost contour line. There are infinitely many cost contours (one for each cost, $C*$). A point on the contour represents the pair of values (D_i, D_I) for which total cost is $C*$, and can be given by

$$D_I = \left[\frac{1}{L_I} \left(\frac{C*}{A} - L_i D_i^b \right) \right]^{1/b} \tag{5.5.20}$$

or for small pipes

$$D_I = \frac{1}{b} \ln \left[\frac{1}{L_I} \left(\frac{C*}{A} - L_i \exp (b \, D_i) \right) \right]. \tag{5.5.21}$$

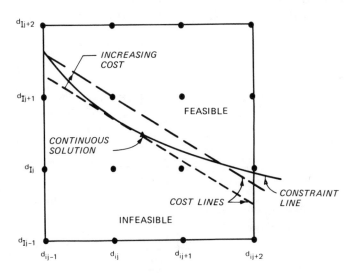

Fig. 5.5. Determining discrete optimum in gravity system.

One need only draw one contour line using either of the above equations and then drawing parallel contours using a pair of right triangles, it is possible to find the contour line closest to the origin (i.e., lowest cost) which intersects a discrete pair of pipe diameters in the feasible region. This pair (d_i, d_I) is the optimal solution. (Note that the contours are actually not straight lines, but their curvature is small over a range of two or three pipe diameters.)

The analytical version of this solution must be solved by trial and error. It involves picking commercially available diameters d_{j-1}, d_j, d_{j+1}, d_{j+2} for $d_j < D_i < d_{j+1}$ and inserting these values into Eq. (5.5.19). The value D_I from Eq. (5.5.19) should be rounded up to the nearest commercial pipe size and the cost of the two pipes determined using

$$C* = A \ (L_i d_i^b + L_I d_I^b)$$ (5.5.22)

or

$$C* = A \ [L_i \exp \ (bd_i) + L_I \exp \ (bd_I).]$$

The pair of pipe sizes with the lowest total cost should be selected.

The procedure for selecting the least cost pipe sizes in a gravity flow system is shown in Fig. 5.6. The steps are described below:

1. Determine water use at each junction for peak and average flow, head loss coefficient (C_w or f) and exponent (m), cost function coefficient (b), initial head (H_0) and minimum head at junctions ($\overline{H}_i$).

2. Calculate flow for peak (Q_i) conditions, and head loss coefficient

$$F_i = K_{2i} L_i (Q_i)^{m-3}$$ (5.5.23)

for each pipe.

3. (a) For large pipes [i.e., cost function (5.3.1)], calculate the diameter of pipe D_1 using

$$D_1 = \left[\frac{\displaystyle\sum_{i=1}^{n} F_i \left(\frac{L_i F_1}{L_1 F_i} \right)^{m/(b+m)}}{H_0 - H_n} \right]^{1/m}$$ (5.5.24)

then select remaining pipes using

$$D_i = D_1 \left(\frac{F_i L_1}{F_1 L_i} \right)^{1/(b+m)}$$ (5.5.25)

GRAVITY LINE

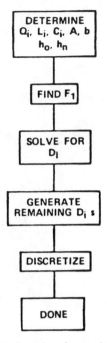

Fig. 5.6. Procedure for gravity line.

(b) For small pipes, determine initial estimates of the D_i's using

$$D_i = \left[\frac{K_2 Q_i^{m-3} \, \Sigma \, L_i}{H_0 - H_n} \right]^{1/m} \tag{5.5.26}$$

where L_i = total length of pipeline, ft (m). Then iteratively solve for D_1 using

$$D_1 = \left[\frac{\sum_{i=1}^{n} F_i[(F_1 L_1 / F_1 L_i) \exp (b \, (D_1 - D_i))]^{-m/(m+1)}}{H_0 - H_n} \right]^{1/m} \tag{5.5.27}$$

and the other D_i's:

$$D_i = \left\{ \frac{F_i L_1}{F_1 L_i} \exp [b(D_1 - D_i)] \right\}^{1/(m+1)} \tag{5.5.28}$$

4. Once the D_i's are known, round pipe diameters to nearest commercially available diameter for all except two diameters (i.e., the two for which the rounding off is greatest). For these (called i and I) it is possible to use the graphical solution described above or a trial-and-error solution. The trail-and-error solution consists of picking several commercial diameters near D_i, say $(d_{j-1}, d_j, d_{j+1}, d_{j+2})$ for which $d_{j-1} < d_j < D_i < d_{j+1} < d_{j+2}$. Calculate the diameter of the other pipe, D_{Ij}, corresponding to these d_j's using

$$D_{Ij} = \left[\frac{F_I}{h* - F_i/d_{ij}^m} \right]^{1/m} \qquad (5.5.29)$$

where

$$h* = H_0 - H_n - \sum_{\substack{k=1 \\ k \neq i, I}}^{n} F_k / d_k^m.$$

When double subscripts are used on d and D in this section, the first refers to the pipe (i or I) while the second refers to the specific discrete value for pipe i. For example, D_{13} is the continuous diameter for pipe I associated with discrete diameter 3 for pipe i.

Round the D_{Ij}'s upward to the next larger diameter (d_{Ij}) and calculate the cost of the two pipes using

$$C* = A[L_i d_{ij}^b + L_I d_{Ij}^b] \qquad (5.5.30)$$

or

$$C* = A[L_i \exp (bd_{ij}) + L_I \exp (bd_{Ij})]. \qquad (5.5.31)$$

Select the pair (d_{ij}, d_{Ij}) that has the lowest value of $C*$.
5. Check to insure that the head is adequate at each junction at peak flow

$$H_I = H_0 - \sum_{i=1}^{I} \frac{F_i}{d_i^m}, \qquad I = 1, 2, \ldots n \qquad (5.5.32)$$

where H_I = head at end of pipe I at peak flow, ft (m).
6. Compare actual head (H_I) to required head $(\bar{H}_I)$ to insure that the heads are reasonable at high flow.

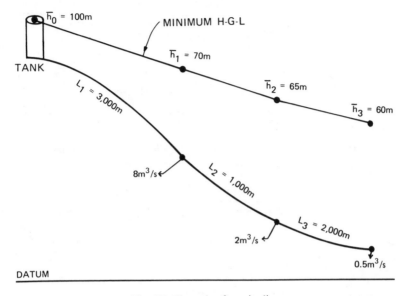

Fig. 5.7. Example of gravity line.

EXAMPLE. Consider the system shown in Fig. 5.7, for which $f = 0.02$, $m = 5$, $b = 1.7$. The flows in each pipe are $Q_1 = 10.5$ m³/s, $Q_2 = 2.5$ m³/s, $Q_3 = 0.5$ m³/s.
The head loss coefficient is

$$F_i = 3.14 \times 10^{13} (0.02) L_1 (Q_i)^2$$

$$F_1 = 2.08 \times 10^{17}, \qquad F_2 = 3.93 \times 10^{15}, \qquad F_3 = 3.14 \times 10^{14}.$$

Since $m/(b + m) = 0.75$, and $1/m = 0.2$. D can be determined as

$$D_1 = \left[\frac{2.08 \times 10^{17} + 3.93 \times 10^{15} \left(\frac{1}{3} \times \frac{2.08}{3.93} \right)^{0.75} + 3.41 \times 10^{14} \left(\frac{2}{3} \times \frac{2.08}{3.14} \right)^{0.75}}{40} \right]^{0.2}$$

$$= 1467 \text{ mm } (57.7 \text{ in.})$$

$$D_2 = \left(\frac{3 \times 3.93 \times 10^{15}}{1 \times 2.08 \times 10^{17}} \right)^{0.15} 1467 = 954 \text{ mm } (37.5 \text{ in.})$$

$$D_3 = \left(\frac{3 \times 3.14 \times 10^{14}}{2 \times 2.08 \times 10^{17}} \right)^{0.15} 1467 = 588 \text{ mm } (23.1 \text{ in.})$$

D_1 should obviously be rounded to 1500 mm (60 in.), for which the head loss is $2.08 \times 10^{17}/1500^5 = 27.4$ m, and $h* = 100 - 60 - 27.4 = 12.6$ m. Since the sizes

**Table 5.3. Trial-and-Error Solution
for Commercial Diameter.**

d_2 (mm)	d_3 (mm)	$C*$ (10^5 $)
750	infeasible	—
900	600	4.85
1050	600	5.57

to which D_2 can be rounded are 750 mm (30 in.), 900 nm (36 in.) or 1050 mm (42 in.), the corresponding values for D_3 are

$$D_3 = \left[\frac{3.14 \times 10^{14}}{12.6 - 3.93 \times 10^{15}/900^5} \right]^{0.2} = 564 \text{ mm}$$

or

$$d_3 = 600 \text{ mm (24 in.)}$$

for which

$$C* = 0.0023 \left[1000 \, (900)^{1.7} + 2000 \, (600)^{1.7} \right] = 4.85 \times 10^5.$$

The results are shown in Table 5.3. It is clear that the optimal diameters are 900 mm (36 in.) and 600 mm (24 in.). These diameters result in heads of

$$H_1 = 100 - \left(\frac{2.08 \times 10^{17}}{(1500)^5} \right) = 72.6 \text{ m}$$

$$H_2 = 72.6 - \left(\frac{3.93 \times 10^{15}}{(900)^5} \right) = 65.9 \text{ m}$$

$$H_3 = 65.9 - \left(\frac{3.14 \times 10^{14}}{(600)^5} \right) = 61.9 \text{ m}.$$

Since the heads are acceptable, the solution is complete.

5.6. RULES OF THUMB AND NOMOGRAMS FOR PIPE SIZING

The techniques for optimal pipe size selection given in the preceding paragraphs are typical of manual methods for pipe size selection, and give optimal results for cases where the design flow in the pipe is known. Even such fairly straightforward procedures, which have been available in the literature for some time, are usually not used by engineers wanting easy answers to the question of which pipes to select. These engineers turn to rules of thumb which are usually couched in terms of velocity or head loss at peak flow. Rule-of-thumb methods are decribed below, as well as some nomograms that are as easy to use as the rules of thumb, but which account for such important factors as energy cost and the relationship between peak and average flows.

The most important variable in pipe size selection is obviously the peak flow to be carried by the pipe. Sizing criteria are often stated in terms of velocity at peak or average flow (e.g., 5 ft/sec at peak flow to 8 ft/sec for fire situations) or in terms of allowable head loss (e.g., 4 ft/1000 ft at peak flow). These criteria can be converted into pipe sizes by solving the continuity equation for pipe diameter to give

$$D = (uQ/V)^{0.5} \qquad (5.6.1)$$

where

Q = peak flow
D = diameter, ft
V = design velocity at peak flow, ft/sec
$u = \begin{cases} 148, & \text{for } Q \text{ in mgd} \\ 0.214, & \text{for } Q \text{ in gpm} \\ 3374, & \text{for } Q \text{ in m}^3/\text{sec.} \end{cases}$

When the rules for pipe sizing are expressed in terms of allowable head loss, the pipe size can be given by

$$D = \left[\frac{uQ}{C(h/L)^{0.54}} \right]^{0.38} \qquad (5.6.2)$$

where

h/L = head loss at flow Q, 1/100
C = Hazen–Williams C factor
$u = \begin{cases} 42.7, & \text{for } Q \text{ in mgd} \\ 0.0615, & \text{for } Q \text{ in gpm} \\ 975, & \text{for } Q \text{ in m}^3/\text{sec} \end{cases}$

While the above rules are attractive because of their simplicity, they are less than ideal methods for selecting pipe size because they ignore many of the factors which determine which pipe size is actually optimal. Table 5.4 gives a list of parameters which can affect pipe size selection and indicates the relative effects on pipe size. With the exception of peak flow, these parameters are not considered in the above rules of thumb.

In general, anything that significantly increases energy costs will cause

Table 5.4. Effect of Parameters on Optimal Pipe Size

As this parameter increases:	Optimal pipe size:
Peak flow	increases
Average flow	increases
Construction price level	decreases
Energy price	increases
Interest rate	decreases
Depth of cover	decreases
Amount of rock excavation	decreases
Wire-to-water efficiency	decreases

optimal pipe size to increase and thus reduce friction head loss. On the other hand anything that increases construction cost disproportionately with respect to energy cost will make smaller pipe sizes more attractive. Pipe length and elevation at the beginning of the pipe do not appear in the table above; however, they enter into the determination of pipe size indirectly insofar as they influence the number of pumping stations required. For example, if a pipe is running downhill, it may be possible to select a sufficiently large pipe size that no pumping is ever required. The cost of that pipeline must be compared with the cost of the optimally sized pumped system to determine which has the lowest life-cycle costs. Similarly, in a long pipeline, a tradeoff must be made between using multiple pumping stations or pipe with a higher pressure rating as opposed to using a larger diameter pipe.

Of the variables listed in Table 5.4, the ones with the greatest effect on pipe size selection are: (1) peak flow, (2) the ratio of peak to average flow ($P:A$), and (3) the relative cost of construction and energy as indicated by

$$C_r = \frac{\text{ENR Construction Cost Index}}{\text{Price of Energy (cents/kwhr)}}. \qquad (5.6.3)$$

The average flow is important, since it is an indicator of the actual flow in the pipe, and it is this actual flow rather than the peak flow that determines the energy cost. In general, it is more convenient to describe average flow by the peak-to-average ratio (which is dimensionless and has the same value regardless of the scale). The price of energy varies widely both with time and geographic location. However, if the energy cost fluctuates in the same manner as the construction price level, there will be no impact on the optimal pipe size. It is only relative changes in energy versus construction costs that make a difference in pipe sizing. This is the reason for defining the cost ratio above.

In order to develop a nomogram for pipe size selection, this author used the

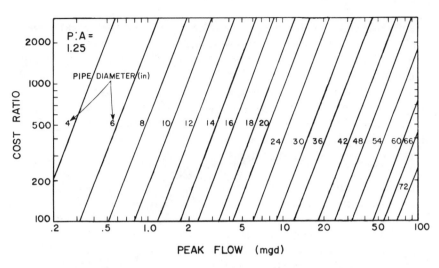

Fig. 5.8. Optimal pipe size for pumped system ($P : A = 1.25$).

MAPS (Methodology for Areawide Planning Studies) computer program* to calculate the optimal pipe size for a large array of combinations of peak flow, peak-to-average ratio, and cost ratio. The MAPS program uses a trial-and-error approach to optimally size individual pipes. The results are shown in Figs. 5.8 and 5.9 for different peak to average ratios. They are based on an interest rate of 8%, O&M labor rate of $10.00/hr, a single pumping station,

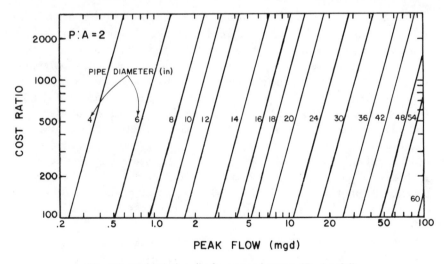

Fig. 5.9. Optimal pipe size for pumped system ($P : A = 2.0$).

*Walski, 1980.

ductile iron pipe, 3 ft of cover, rectangular trench, no rock excavation by blasting, no wet well at the pumping station, horizontally mounted centrifigal pumps, wire-to-water efficiency of 60%, and two gate valves along the pipe. The nomograms can be extrapolated to situations which do not vary greatly from this base condition since the effects of these parameters on optimal pipe size are not great (even though some greatly affect cost). The engineer is, however, cautioned about applying the nomograms to situations for which the conditions given above are not valid.

To use the pipe sizing nomograms:

1. Calculate $P:A$.
2. Select correct figure.
3. Calculate C_r.
4. Draw vertical line through the peak flow.
5. Draw horizontal line through the cost ratio.
6. The lines will intersect in region corresponding to optimal pipe size.

Note that the boundary lines between pipe sizes are less vertical for $P:A = 1.25$ than for $P:A = 2.0$ because the energy costs and hence C_r are less important when the average flow is relatively low. As $P:A$ increases (e.g., lines sized primarily for rare events such as fire flow), the effect of the cost ratio becomes negligible and pipe size can be determined primarily based on the head and flow required at peak flow. This is often the case in distribution lines sized strictly for fire flow.

The form of Figs. 5.8 and 5.9 was selected because cost ratio and peak flow are continuous variables and pipe diameter is a discrete variable. In cases for which the diameter can be treated as a continuous variable, or when English nominal pipe diameters are inappropriate, it is more convenient to select pipe sizes by first determining the optimal velocity or head loss at peak flow and then converting the results to pipe diameter using Eqs. (5.6.1) and (5.6.2). Optimal velocities and head losses for pipe sizing are shown in Figs. 5.10 and 5.11 for several combinations of peak-to-average ratio and cost ratio. The figures show that basing pipe sizing decisions on a constant design velocity, or head loss, can lead to some inefficient pipe size selections. For example, suppose an engineer sizes pipes based on a velocity of 6 ft/sec, which is the correct velocity for $P:A = 2.$, $C_r = 700$ and $Q = 4.2$ mgd. For a flow of 20 mgd with the same $P:A$ and C_r, the optimal velocity is 7 ft/sec.

EXAMPLE. Consider the problem in which an engineer must select a pipe size to produce cost estimates in a planning study, given the following data:

Peak flow: 10 mgd
Average flow: 8 mgd

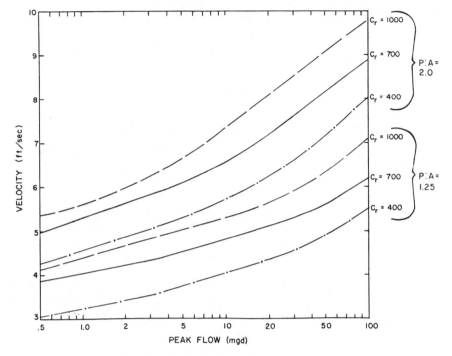

Fig. 5.10. Optimal velocity for pumped system.

Energy costs: 8 cents/kwhr
ENR Index: 4000
One pump station required.

The engineer should first calculate $P:A$ as 1.25, which means Fig. 5.8 should be used. The C_r can be calculated as $4000/8 = 500$. Knowing this and the peak flow, it is easy to read the diameter of 24 in. for the solution.

Another approach is to calculate $P:A$ and C_r as above, then using Fig. 5.10, determine the velocity at peak flow to be 4.2 ft/sec. Using Eq. (5.6.1), the diameter is given as 25.9 in., which can be rounded to 24 in.

The simple procedure given in this section should not be used to select pipe sizes without checking to be certain that the overall system will meet head and flow requirements. For complicated systems, this is best done with a simulation model.

Any pipe size selection method must be tempered by consideration of waterhammer. Since there is roughly a surge of 100 ft of pressure for every 1 ft/sec of velocity change (100 m for every 1 m/sec), engineers generally will prefer to avoid velocities where large surges are possible because of the cost of

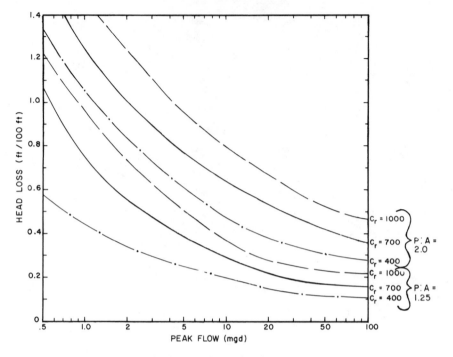

Fig. 5.11. Optimal hydraulic gradient for pumped system.

surge control equipment, and simply to provide an extra safety factor. If a pipeline is long enough to require surge protection, the cost of protection is only slightly dependent on the magnitude of the surge. Nevertheless, some engineers use rules-of-thumb such as 5 ft/sec at peak flow as an upper limit on velocity.

5.7. MATHEMATICAL PROGRAMMING APPROACH TO PIPE SIZING

This chapter has focused primarily on manual methods for pipe sizing for systems that can be treated as non-looped. This is not to imply that looped systems are not important—on the contrary, almost all major water distribution systems can be considered as looped. The problem is that there are no straightforward manual procedures for optimally sizing looped systems. Even computerized methods for optimizing looped systems leave much to be desired at this time.

The engineer is therefore forced to make pipe sizing decisions based on trail-and-error methods, coupled with rules of thumb similar to those presented in the previous section. The reason for this limitation in analytical tools lies in the difficulty of specifying redundancy to optimization techniques. Given the fact that water must be moved from a source to a collection of use points in a distribution system, both manual and computerized techniques can give optimal or nearly optimal values for pipe size. The optimal system, however, would always be a non-looped system, as redundancy results in extra cost for the same carrying capacity. Redundancy allows more than one path to reach any point in the system.

Redundancy can be provided by forcing certain pipes to be included in the system at a minimum size regardless of the cost, or by specifying that at least two pipes must be connected to each node. These are fairly arbitrary criteria, but they do force the system to be looped. The problem of specifying design criteria for distribution systems is discussed in more detail in the following section.

Once the problem of how to specify redundancy to an optimization model is solved, selecting optimal pipe sizes can begin. However, it is no longer possible to know the flow in each pipe beforehand, as was true for the earlier methods. The flow in each pipe now depends on the diameter, which in turn depends on the flow. The obvious way to get around this problem is to use an iterative solution in which the flow (or head) distribution is specified; then the pipe sizes are calculated using an optimization scheme. The flows are then adjusted using some kind of gradient search technique and the procedure is repeated until a better solution cannot be reached. Alperovits and Shamir (1977), Quidry, Brill, and Liebman (1981), and Morgan and Goulter (1982) have developed methods to solve this type of problem. Kettler and Goulter (1983) have developed an optimization procedure which explicitly considers reliability in branched systems.

Others have attempted to impose criteria such as a constant hydraulic grade line on the system to make the problem workable. Gessler (1982) and Gessler and Walski (1984) have approached the problem from the other direction and proposed a method that seeks to take advantage of the availability of pipes only in discrete sizes. This model essentially performs a trial-and-error solution for an array of possible diameters for each pipe as shown in Fig. 5.12. This method works well when the number of pipes under consideration is not large, which is usually the case for expansion of existing systems.

In general, the cost of setting up and running these optimization programs, is fairly large, and a true optimal solution cannot be guaranteed. However, it is only a matter of time before these techniques become developed to the point where they are as commonly used as the Hardy–Cross and Newton–Raphson methods are used today for hydraulic problems.

5.8. DESIGN CRITERIA

Since water distribution systems are found almost everywhere that more than one or two homes are grouped together, one would think that design criteria for these systems would be fairly standardized. Such criteria for pipe dimen-

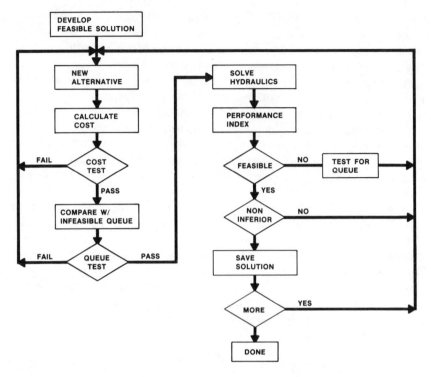

Fig. 5.12. Flow chart for discrete pipe sizing method.

sions and structural integrity have been developed by the American Water Works Association, but in the area of pipe sizing, current criteria appear to be rather arbitrary. There are as many sets of criteria as there are agencies to promulgate them.

Most criteria are actually performance criteria couched in terms of pressure, such as: 60 psi or 40 psi minimum pressure except for 20 psi during fires. Some specify the flow that the system must carry as a function of the number of persons or households served. Little mention is made of whether the pressure criteria apply when a pipe or pump may be out of service. Some rules specify a minimum pipe size (usually 2 in.) and a minimum size for a system with fire protection (usually 6 in.).

The Insurance Service Office, which publishes the Fire Suppression Rating Schedule, has a system for assigning the Public Protection Classification of systems based on the fire department, water supply and means for handling fire calls. Water supply is judged by comparing the Needed Fire Flow (NFF) with the (1) Supply Works Capacity, (2) Main Capacity, and (3) Hydrant Distribution. The capability of the water system is determined, for each test location, by the least of the three items listed above. The Main Capacity is the amount of water the system can deliver at the test location at a pressure of 20 psi. The guidance does not state at what time of day the tests should be run, what the boundary conditions should be (i.e., what pumps should be running and what the water level in the tanks should be), or how the test locations should be selected. Furthermore, the formula for calculating flow at 20 psi is conservative for systems with pressure reducing valves and a large non-fire flow.

With the criteria from the regulatory agencies and the Insurance Services Office, the engineer has a great deal of leeway in designing safety factors into the system. Some would say that the system should be able to meet pressure criteria with any one (or two) pipe segments valved off, or with all of the tanks empty. Others would claim that the pressure criteria contain an allowance for the fact that the system will not always perform as designed.

Testing the system with a network model or calculating optimal pipe sizes for different modes of failure is a desirable way to analyze a system, but the number of combinations of failure modes is prohibitive. deNeufville, Schaake, and Stafford (1971) discuss the fact that using simple, single-objective design criterion does not result in the best design from a cost or reliability standpoint. They argue in favor of developing indicators of effectiveness of various measures as a means of selecting the design that provides the best system at the least cost.

At present, the engineer has a good deal of freedom to choose the amount of overdesign and redundancy to be built into a system. Shamir and Howard (1981) have described techniques to quantify the reliability of a distribution system but depend heavily on how reliability is defined (e.g., volume of shortfall, duration of shortfall, combination of pressure and volume shortfall). Since it is unreasonable to expect a system never to fail, and it is prohibitively expensive to design around all such failures, some acceptable risk for water system design should be determined and built into design criteria. Unfortunately, such criterea are presently not available.

REVIEW QUESTIONS

1. Why aren't trial-and-error solutions likely to give you optimal pipe sizes for complex distribution systems?
2. Why must gravity flow and pumped system problems be treated differently in pipe sizing?
3. Why must loops be forced into optimal pipe sizing problems?
4. When and where would you conduct a fire flow test on a system if you wanted the system to look good? look bad?
5. Why aren't mathematical programming techniques suitable for manual calculations?
6. What problems result in using a pipe sizing criterion of 5 ft/sec for all distribution systems?

7. Why is it difficult to quantitatively account for reliability in water distribution systems?
8. Why do energy costs become important in calculating pipe size when the average flow is close to peak flow?
9. What problems are introduced in pipe sizing problems by the fact that pipes are only available in specific nominal diameters?
10. Why should pressure not drop below 20 psi for fire fighting?

PROBLEMS

1. Calculate the energy and construction cost for 14, 16, 18, 20, and 24 in. pipes carrying 2 mgd peak and 1.6 mgd average flow from a pump, if the cost of energy is 12 cents/kwhr, pump wire-to-water efficiency is 70%, the interest rate is 10% and the Hazen–Williams C is 120. Use a capital cost function of $1.2D^{1.6}$. Plot the present worth of cost per foot of pipe versus diameter for each diameter. Then calculate the optimal pipe size using the method given in Section 5.4 and the nomogram in Section 5.6. Compare the results.
2. Given the gravity flow system in Fig. 5.1p, find the optimal pipe sizes using $C =$

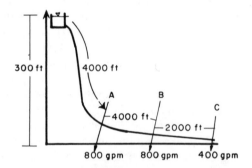

Fig. 5.1p. System for Problem 2.

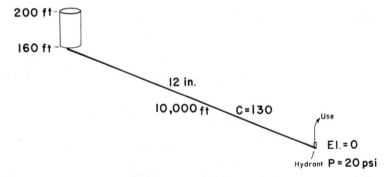

Fig. 5.2p. System for Problem 3.

120, cost $= 1.0D^{1.4}$, and $m = 4.85$. The heads required at junctions A, B, and C are 160 ft, 160 ft, and 150 ft, respectively. (*Solutions:* 12, 10, 6 in.)

3. Given the one-pipe system shown in Fig. 5.2p, use the Hazen–Williams equation to calculate the fire flow that can be delivered to the end of the pipe at 20 psi for the following cases: (a) tank full, consumptive use at the end of the pipe $= 500$ gpm; (b) tank full, consumptive use $= 2000$ gpm; (c) tank empty, consumptive use $= 500$ gpm; (d) tank empty, consumptive use $= 2000$ gpm. (*Solutions:* 1730, 230, 2124, and 624 gpm)

REFERENCES

Alperovits, E., and U. Shamir, 1977, "Design of Optimal Water Distribution Systems," *Water Resources Research*, Vol. 13, No. 6, p. 885.

Cowan, J., 1971, "Checking Trunk Main Designs for Cost-Effectiveness," *Water and Water Engineering*, Vol. 75, No. 908, p. 385.

Dances, L., 1977, "Sizing Force Mains for Economy," *Water and Sewage Works*, 1977 Reference Issue, p. R-127.

Deb, A.K., 1973, "Least Cost Design of Water Main in Series," *J. ASCE Environmental Engineering Division*, Vol 99, No. EE3, p. 405.

Deb, A.K., 1976, "Optimization of Water Distribution Network Systems," *J. ASCE Environmental Engineering Division*, Vol 102, No. EE4, p. 837.

Featherstone, 1983, R.E., and K.K. El-Jumaily, "Optimal Diameter Selection for Pipe Networks," *ASCE Journal of Hydraulic Engineering*, Vol. 109, No. 2, p. 221.

Gessler, J., 1982, "Optimization of Pipe Networks," *1982 International Symposium on Urban Hydrology, Hydraulics and Sediment Control*, Lexington, KY, p. 195.

Gessler, J. and T.M. Walski, 1984, "Selecting Optimal Strategy for Distribution System Expansion and Reinforcement, *Urban Water 84; A Time for Renewal*, Baltimore, MD.

Kettler, A.J., and I.C. Goulter, 1983, "Reliability Considerations in the Least Cost Design of Looped Water Distribution Systems," *1983 International Symposium on Urban Hydrology, Hydraulics and Sediment Control*, Lexington, KY, p. 305.

Morgan, D.R., and I.C. Goulter, 1982, "Least Cost Layout and Design of Looped Water Distribution Systems," *1982 International Symposium on Urban Hydrology, Hydraulics and Sediment Control*, Lexington, KY, p. 65.

Nolte, C.B., 1979, *Optimal Pipe Size Selection*, Gulf Publishing Co., Houston, TX.

Osborne, J.M., and L.D. James, 1973, "Marginal Economics Applied to Pipeline Design," *J. ASCE Transportation Division*, Vol. 99, No. 3, p. 637.

Quidry, G.E., E.D. Brill, and J.C. Liebman, 1981, "Optimization of Looped Water Distribution Systems," *J. ASCE Environmental Engineering Division*, Vol. 107, No. EE4, p. 665.

Rasmussen, H.J., 1976, "Simplified Optimization of Water Supply Systems," *J. ASCE Environmental Engineering Division*, Vol. 102, No. EE2, p. 313.

Rowell, W.F. and J.W. Barnes, 1982, "Obtaining Layout of Water Distribution Systems," *J. ASCE Hydraulics Division*, Vol. 108, No. HY1, p. 137.

Schaake, J.C., and D. Lai, 1969, "Linear Programming and Dynamic Programming Application to Water Distribution Network Design," MIT Hydrodynamics Lab Report 116, Cambridge, Mass.

Shamir, U., 1974, "Optimal Design and Operation of Water Distribution Systems," *Water Resources Research*, Vol. 10, No. 1, p. 27.

Shamir, U., 1979, "Optimization in Water Distribution Systems Engineering," *Mathematical Programming*, No. 11, p. 65.

Stephenson, D., 1976, *Pipeline Design for Water Engineers*, Elsevier Scientific Publ.

Swamee, P.K., V. Kumar, and P. Khanna, 1973, "Optimization of Dead End Water Distribution Systems," *J. ASCE Environmental Engineering Division*, Vol. 99, No. EE2, p. 123.

Walski, T.M., 1982, "Cost Savings Through Pipe Design for Energy Efficiency," *AWWA National Conference*, Miami, FL.

Walski, T.M., 1980, "Energy Costs: A New Factor in Pipe Size Selection," *J. AWWA*, Vol. 72, No. 6, p. 325.

Watanatada, T., 1973, "Least-Cost Design of Water Distribution System," *J. ASCE Hydraulics Division*, Vol. 99, No. HY9, p. 1497.

6 | Providing and Restoring Carrying Capacity

6.1. INTRODUCTION

Water mains gradually lose carrying capacity over time. In particular, unprotected iron and steel pipes carrying corrosive water can lose over half of their carrying capacity (as measured by the Hazen–Williams constant) within 30 years of installation. In recent times, metal pipes lined with cement mortar, bituminous enamel, or plastic to inhibit such corrosion have become very popular for drinking water application, but there are still millions of miles of unlined metallic pipe in use today.

Corrosion affects the carrying capacity of pipes by pulling iron out of the pipe to form tubercles and pits on the interior surface, as shown in Fig. 6.1. Cleaning and lining can restore the original carrying capacity by removing the tuberculation and preventing its reoccurence, Figure 6.2 shows a pipe that has been cleaned and lined.

Carrying capacity can also be lost by the deposition of excess calcium carbonate scale on the inside of pipe because the water is supersatured with calcium. This is especially true with systems fed by wells where it would be prohibitively expensive to treat the water at the source to produce a nonscaling water.

Carrying capacity can also be reduced by slime growths (which are more of a problem in nonpotable water pipes), deposit of sediment, and excess lead jointing material which forms lumps on the inside of pipes at joints. While only metal pipes are subject to tuberculation and lead joints are no longer installed on new pipes, carrying capacity of any type of pipe can be reduced by the other mechanisms.

If the scale on the pipe consists of calcium carbonate, a thin layer can be left behind to protect the pipe. If the original tubercles were caused by corrosion, then pipe must be lined, as otherwise the carrying capacity will be lost fairly rapidly. In any case the benefits of cleaning and lining can be maintained indefinitely if the water is properly stabilized at the treatment plant to prevent scaling or corrosion.

Fig. 6.1. Pipe with tuberculation. [Courtesy Ameron.]

In the following sections, the loss of carrying capacity is described and a detailed discussion of the process for cleaning and lining mains is presented. Methods for deciding which lines can be cleaned and lined economically depend on whether the pipe has adequate capacity and whether the alternative is installation of parallel piping or simply greater expenditure for pumping energy. Once the costs for cleaning and lining are described, criteria for deciding when it is economical are presented.

6.2. LOSS OF CARRYING CAPACITY

Lost carrying capacity is generally described in terms of the Hazen–Williams C factor. New pipes generally have a C of about 130. This decreases rapidly for unlined metal pipes, with the rate depending on the corrosivity of water. Table 2.3 shows how quickly carrying capacity can decline. The roughness elements grow at a roughly linear rate with time and, since the C factor is a logarithmic function of roughness element size, C decreases logrithmically, as shown in Fig. 6.3. The rate of decrease is more dramatic for smaller pipes since the roughness elements occupy a greater fraction of the cross-sectional area than the same elements would occupy in a large main.

Fig. 6.2. Pipe after rehabilitation. [Courtesy Ameron.]

Cleaning the tuberculation from the pipe would seem sufficient to restore the carrying capacity for some time, but experience has shown that a cleaned pipe loses carrying capacity even faster than a new pipe. The exception is the case of a previously lined main where a thin layer of calcium carbonate will protect the pipe if the pipe is not scraped to bare metal. It is therefore usually necessary to line a pipe in order to maintain the high C factor that was achieved by cleaning. After cleaning and lining, C factors are usually on the order of 120 for small mains to as high as 140 for larger mains.

The rate at which pipe roughness increases can be minimized by chemically stabilizing the water at the treatment plant. This is usually done by addition of lime and possibly carbon dioxide to produce a water that is neither aggressive nor scaling. Polyphosphate compounds are also used to protect the interior of water mains but are usually economical only for small treatment plants.

There are numerous indicators of the chemical stability of water. The Langlier index and the Ryznar index essentially compare the saturation pH of the water with the actual pH to determine if the water is saturated. Caldwell–Lawrence diagrams provide a graphical solution to this problem. Merril, Sanks, and Spring (1978) describe these methods for determining the stability

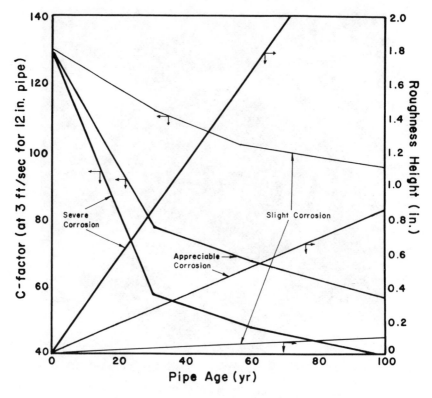

Fig. 6.3. Increase in pipe roughness with time.

of water with respect to calcium carbonate. Morgan, Walski, and Corey (1984) developed graphical and microcomputer methods for determining lime and carbon dioxide doses to produce a stable water.

Stabilization (also called conditioning) of water can slow the tuberculation or scaling process but cannot restore lost capacity. This can only be done by cleaning. However once a pipe is rehabilitated, chemical stabilization of the water will help to assure that carrying capacity will be maintained indefinitely.

6.3. CLEANING AND LINING PROCESS

Cleaning and lining water mains is a major operation, but if done properly it can indefinitely restore the carrying capacity of water mains. It can be accomplished with a minimal amount of interference with traffic and interruption of service. The individual steps in the process are described in the following paragraphs. Fig. 6.4 shows the cleaning and lining operation for

For Pipelines 4 Inches (100 mm) Through 36 Inches (914 mm) in Diameter

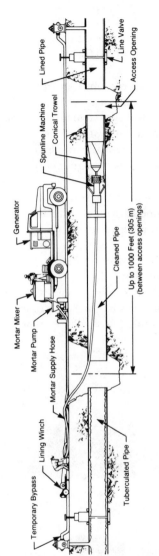

For Larger Pipelines to 264 Inches (6.7 m) in Diameter

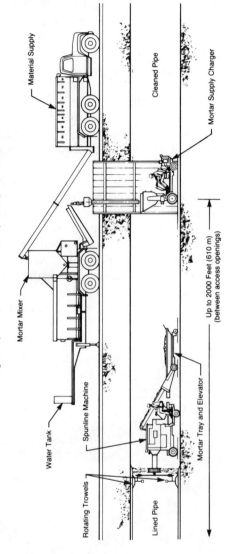

Fig. 6.4. Cleaning and lining process. [Courtesy Ameron.]

small and large diameter mains. Figs. 6.5 and 6.6 show some of the machines used to place the cement mortar in pipes. For small pipes, the typical section to be lined is 500 ft while for larger mains (greater than 40 in.) the sections can be as long as 1500 ft.

Once the pipe section has been selected, bypass piping must be installed to provide service to customers along the line and, in some cases when the system is not highly looped, serve other neighborhoods downstream. The bypass piping is usually connected to hydrants beyond the ends of the section being cleaned and lined. Temporary service connections are made at meter boxes, hose bibs, or any other location at which customer lines can be connected with the bypass lines.

Openings are made at each end of the pipe by digging down to the pipe, shoring the trench, and removing a section of pipe so that entry can be gained to the inside of the pipe. In some large mains it is only necessary to provide access at one end of the pipe. The section of the pipe that is removed for access is cleaned and lined by hand outside the trench. That section of pipe can be easily examined for structural integrity, external corrosion, etc.

Scrapers are then passed through the line to clean the pipe. They are usually pulled mechanically by winches, although some can be forced through the lines hydraulically if adequate pressure is available. For very large pipes the

Fig. 6.5. Small mortar machine with trowel. [Courtesy Raymond International.]

Fig. 6.6. Large mortar machine. [Courtesy Raymond International.]

scrapers are usually electrically driven and manually operated. Usually several passes are required to clean the pipe.

Debris is flushed from the pipe and a squeegee assembly is pulled through the line to remove the last of the debris and water. Any major joint leaks are repaired at this time.

Next a mortar spraying machine is pulled through the pipe followed by a series of trowels. The pipe is then closed with bulkheads to maintain optimal moisture conditions for curing. Before the cement dries compressed air is blown through corporation cocks and other openings to insure that they do not become clogged with cement. Valves should also be checked to insure that the cement does not interfere with their operation.

The lines are then inspected, flushed, and disinfected and, the sections that were removed to provide access are replaced. Once service is reestablished, the bypass piping is removed and the access holes are repaved. Standards for cement-mortar lining of iron pipes are given in AWWA C602-76.

Cleaning and lining equipment cannot pass through butterfly valves, check valves, or globe valves, and lining equipment does not work well at sharp bends. Generally these valves and bends are removed during the cleaning and lining operation so that they can receive maintenance, be lined by hand, and not interfere with the rest of the operation.

While cable-pulled scrapers are the most effective method for pipe cleaning, in many cases hydraulically propelled foam pigs can be used. These pigs can be launched from hydrants or specially installed pipe sections or fittings. Usually several pigs are required to clean a pipe. First a soft, bare pig is passed through the pipe to insure that there are no obstructions and to assist in estimating the minimum internal diameter. Next, progressively larger pigs with harder coatings are passed through the pipe to clean the pipe. Pipes that have been cleaned using pigs are generally not cement mortar lined.

While cement mortar lining is usually the most effective and economical method for lining pipes, it will only seal tiny leaks in the pipe. Where leakage is a problem, pipes can be lined with processes called sliplining or insituform. Sliplining consists essentially of inserting a thin-walled plastic pipe into the old pipe. While it prevents leaks, it reduces the diameter of the lined pipe by slightly more than twice the thickness of the liner. Therefore, the C-factor of the lined pipe can be given by

$$C_e = C_L (D_L/D)^{2.63} \qquad (6.3.1)$$

where C_e = effective C-factor for lined pipe based on nominal diameter
C_L = C-factor of lining based on actual diameter
D_L = diameter after sliplining
D = diameter used to calculate C

Sliplining is comparable economically with cement mortar lining only for small diameters, and it requires a larger excavation than cement mortar lining. It is primarily attractive where leakage is a problem.

Insituform is a process that was developed for patching leaking sewers but can be used in water lines for moderate sized holes. In this process, a chemically impregnated fabric is forced into the pipe and water forces the fabric to conform to the shape of the pipe. The water is heated to cure the resin and holes are cut at the services. This process is still fairly expensive in comparison with cement mortar lining.

6.4.   ECONOMIC EVALUATION OF CLEANING AND LINING

Cleaning and lining is just one of several alternatives that a utility has when it notices low pressures or poor results from fire flow tests. The others are: (1) learn to live with low pressure; (2) adjust pump station operation or add more pumping head to provide the desired pressure; (3) install parallel piping; or (4) install elevated storage or hydropneumatic tanks. The first alternative should be unacceptable for a well-run utility. The fourth is only economical for an isolated area with many small pipes and no existing elevated storage. This is not usually the case in systems old enough to have problems with lost carrying capacity. Quantitative methods to decide among the remaining alternatives are presented in the following sections. In general, cleaning and lining is most economical for large mains which have very low C factors initially and have the capacity and structural integrity to meet future demands if cleaned and lined.

When faced with a pressure problem, a utility's first response should be to monitor pressure and conduct head loss tests to determine pipe roughness, since the decision to clean and line is highly dependent on the initial pipe roughness. These tests are described in Chapter 8.

The rules for cleaning and lining presented in later sections are based on the simple criterion that cleaning and lining should be done if it is less expensive than other measures. The data required to perform these comparisons include interest rates, design flow, price of energy, and expected final C factor. Typical numbers will be used in examples, but these values should not be applied generally because they are highly project specific.

There are some intangible factors that must be considered in deciding to clean and line pipes. Cleaning and lining can eliminate red water problems caused by the pipe which is cleaned and lined. This problem will not be totally eliminated, however, until all lines are rehabilitated. This is seldom economically justifiable. Lining can also reduce leakage by blocking small holes, especially at joints. When compared with installation of a new main, cleaning and lining results in less disruption of traffic, since entry to the pipe is at roughly 500-ft intervals. However, when compared to additional pumping, cleaning and lining results in more disruption. It is difficult to quantify these factors, so a considerable amount of judgment is required.

The key to deciding if the best alternative to cleaning and lining is more pumping or installation of parallel pipe lies in determining if the pipe could carry adequate flow with acceptable head loss without cleaning and lining. For pipes to be considered for cleaning and lining the following criterion (head loss greater than allowable head loss) must be met:

$$\frac{KL}{D^m}\left(\frac{Q}{C}\right)^n > h \qquad (6.4.1)$$

where

L = length of pipe under consideration, L
K = constant in Hazen-Williams equation
    (= 10.4 when Q in gpm and D in in.)
D = diameter of pipe, I

Q = design flow, L^3/T
C = Hazen–Williams C before cleaning
h = allowable head loss at design flow, L
n = exponent on flow in head loss equation
m = exponent on diameter in head loss equation.

It is helpful to divide Eq. (6.4.1) by the quantity on the left-hand side to yield a dimensionless number which serves as an indicator of the capacity of the existing pipe at design flow. This dimensionless head loss (Y) is given by

$$Y = \frac{h}{L}\left(\frac{C}{Q}\right)^n \frac{D^m}{K}. \tag{6.4.2}$$

If $Y > 1$, the pipe can carry the design flow with acceptable head loss. Cleaning and lining may not be required, but may still be justified based on savings in energy cost. This case is considered in Section 6.6. If $Y < 1$, something must be done to provide additional carrying capacity. In this case it is more appropriate to compare cleaning and lining with the cost of parallel pipe which would give the same results. This analysis is described in Section 6.7.

The results of the analyses are highly dependent on the parameter Y, which is very sensitive to the maximum allowable head loss h (or gradient h/L) and the design flow rate Q. Both values should be selected based on future head and flow requirements, as opposed to current values, and should take into account overall expansion plans for the system. For example, Y can be calculated based on Q required for fire flow and h to maintain 20 psi at some critical point in the system; or Q can be peak hourly use while h is the loss that will result in 40 psi. The Y value to be used is the lower of the values calculated.

EXAMPLE. Determine Y for a 6000-ft-long, 16-in.-diameter pipe which must carry 2500 gpm and yield 40 psi at the downstream end during normal high-use conditions, and a total of 4000 gpm during high use with a fire flow with a pressure of 20 psi. The downstream end is at elevation 300 ft while the head at the source is 440 ft and the C factor is 80.

The allowable head loss must first be determined at both high and low flows:

$$h(2500) = 440 - [300 + (40)(2.31)] = 47.6 \text{ ft}$$

$$h(4000) = 440 - [300 + (20)(2.31)] = 93.8 \text{ ft.}$$

The corresponding values of Y are

$$Y = \frac{47.6}{6000} \left(\frac{80}{2500}\right)^{1.85} \frac{16^{4.86}}{10.4} = 0.93$$

$$Y = \frac{93.8}{6000} \left(\frac{80}{4000}\right)^{1.85} \frac{16^{4.86}}{10.4} = 0.77.$$

The fire flow condition is therefore the critical condition on which cleaning and lining decisions should be based.

In applying the procedures described in this section to a real system, it is necessary to carefully select which mains are to be considered for the analysis. Mains which radiate out from the treatment plant/pumping station should be selected, rather than lines perpendicular to that direction (i.e., lines used to close loops in the system). Attention should be paid to the lines which have the highest hydraulic gradient and/or must support the largest growth in water use, as this is where the greatest benefits will be realized. If only a portion of a long line is to be cleaned and lined or paralleled, that portion should be on the upstream end of the line. In many cases, however, the actual selection will have to be based on ease of access.

In the following sections, the costs of cleaning and lining are discussed. This is followed by a presentation of some techniques for deciding when to clean and line pipes as opposed to when to provide additional pumping or parallel piping.

6.5. COSTS

The cost to clean and line a pipe depends on a fairly large array of variables, which include: (1) pipe diameter and type, (2) overall size of the job; (3) geographic location of system; (4) pipe alignment; (5) size, location, and type of valves; (6) depth of cover and paving type; (7) access to site; (8) lengths which can be accessed at one time; (9) bypass requirements; (10) amount of work done by utility; (11) time of year; and (12) duration of job.

With so many factors affecting cost, it is easy to understand why the cost per foot of pipe can vary from $10 to $50. As a rough approximation the following equation can be used to estimate cleaning and lining costs although there will be considerable variation about this function because of number of excavations, type of paving, length of temporary services, number of valve replacements and pipe obstructions, and total size of job.

$$\text{Cost} = 9.27 D^{0.30}$$

where
$$D = \text{diameter, in.}$$
$$\text{Cost} = \text{cost per foot in 1982 dollars.}$$

For newer cement-lined mains, where the problem is either scaling or growth on the walls, pipes can be cleaned at a lower cost than cleaning and lining, as long as the existing lining is not damaged during the cleaning process.

6.6. COMPARISON WITH ENERGY COST

If the pipe has carrying capacity adequate to provide design flow with acceptable head loss (i.e., $Y > 1$), cleaning and lining may still be justified by savings in energy and equipment for pumping. The criterion for selecting pipes for cleaning and lining is that the pipe should be cleaned and lined if the cost for that operation is less than the present worth of additional energy and pumping costs that would be incurred if the pipe were left alone:

$$c_r < (c_p - c_{cp}) + (c_e - c_{ce})\text{spwf} \qquad (6.6.1)$$

where

c_r = cost of cleaning and lining, \$/ft
c_p = cost of new pumping equipment without cleaning and lining, \$/ft
c_{cp} = cost of new pumping equipment with cleaning and lining, \$/ft
c_e = cost of energy without cleaning and lining, \$/ft/yr
c_{ec} = cost of energy with cleaning and lining, \$/ft/yr
spwf = series present worth factor

(Note that in this chapter a subscript c will refer to the value of a given parameter after cleaning and lining.) When the inequality in (6.6.1) is true, it is economical to clean and line. The series present worth factor should be based on the time over which saving in energy cost will occur. If the pipe is not adequately lined after being cleaned, this may be as short as 5 years, but in most cases the savings will be realized indefinitely. For current interest rates the series present worth factor is insensitive to time for periods greater than about 15 years, so a 20-year planning horizon can be used.

In cases where no additional pumping equipment is needed to serve the main, the terms relating to savings in pumping equipment should be eliminated and the comparison should be made between energy cost and the cost of cleaning and lining. In the following paragraphs, formulas for determining the savings are developed.

The energy lost per length of pipe can be given by the Hazen–Williams equation as

$$h_a = \frac{K}{D^m} \left(\frac{Q_a}{C} \right)^n \qquad (6.6.2)$$

where

h_a = head loss per length of pipe at average flow, ft/ft

K = constant in Hazen–Williams equation (10.4 for Q in qpm and D in in.)

Q_a = average flow, gpm

C = Hazen–Williams C factor

D = pipe diameter, in.

m = exponent on diameter in head loss equation (4.86 for Hazen–Williams)

n = exponent on flow in head loss equation (1.85 for Hazen–Williams).

(Note that in this section h refers to head loss per unit length of pipe, since the economic comparisons are made based on cost per unit length. This is different from use of h as head loss in length units in other sections of this book.) The annual cost of energy to move the water through one foot of pipe depends on the average flow, head loss at average flow and the price of energy, and can be given by

$$c_e = 0.0164 Q_a h_a P / e \qquad (6.6.3)$$

where

P = price of energy, cents/kwhr

e = wire-to-water efficiency of pumps.

The difference between the cost of energy with and without cleaning and lining can be determined by substituting the head required with and without the work into Eq. (6.6.3) and finding the difference in annual energy cost per foot as

$$c_e - c_{ec} = \frac{0.1711 Q_a^{n+1} P}{eD^m}\left(\frac{1}{C^n} - \frac{1}{C_c^n}\right).$$ (6.6.4)

In cases where no new pumping equipment is required for the system, the engineer need only compare the left side of Eq. (6.6.4), multiplied by the series present worth factor, with the cost to rehabilitate a foot of pipe. Eq. (6.6.4) is only an approximation to the energy savings. A more precise answer can be obtained by realizing that cleaning and lining enables the utility to run fewer pumps or run the same pumps for a shorter period during the day. Such an analysis would require knowing a great deal about the system head curves and pump head curves, and in most cases would not produce a significantly different result.

As a pipe loses carrying capacity, or if additional pumping capacity is needed to meet increased demands, the head provided by new pumping equipment must be higher if the pipe is not cleaned and lined. The individual item in a new or upgraded pumping station that is most affected by the loss of carrying capacity is the cost of the driver for the pump. The extra head that must be provided by the pump per foot of pipe length is given by

$$h_p - h_{pc} = \frac{KQ_p^n}{D^m}\left(\frac{1}{C^n} - \frac{1}{C_c^n}\right).$$ (6.6.5)

where

Q_p = peak design flow, gpm
h_p = head provided by pump per foot of pipe length, ft/ft.

The cost savings realized by saving a foot of pumping head depend strongly on the flow provided by the pumps. The incremental cost per additional foot of head, holding the flow constant, is roughly independent of the head (i.e., adding 1 ft to a station rated at 100 ft affects cost just as much as adding a foot to a station rated at 400 ft). Since this cost depends on the flow it can be approximated by the function

$$F = 0.65 Q_p^{0.77}$$ (6.6.6)

where F = cost for an additional foot of head at the pump station, $/ft. The results given by Eq. (6.6.6) are based on costs prepared using the MAPS (Methodology for Areawide Planning Studies) computer program and are expressed in 1982 dollars. These costs should be viewed as very rough estimates, and are acceptable for screening purposes only.

The savings in pumping equipment can therefore be given by

$$c_p - c_{pc} = F(\text{pwf})(h_p - h_{pc}) \qquad (6.6.7)$$

where pwf = present worth factor. The present worth factor reflects the fact that the additional pumping equipment may not be required for some time. If the pumping equipment is required in the current year, then pwf = 1. pwf decreases as the need for equipment occurs further into the future.

Substituting for F and h in Eq. (6.6.7) gives

$$c_p - c_{pc} = \frac{6.76\, Q_p^{n+0.77}}{D^m} \left(\frac{1}{C^n} - \frac{1}{C_c^n} \right)(\text{pwf}). \qquad (6.6.8)$$

The total present worth of savings that can be realized by cleaning and lining can be determined by summing the savings in energy and equipment to give

$$c_R = 10.4\, C^* D^{-m}[100\, Q_p^{n+.77}(\text{pwf}) + Q_a^{n+1} P(\text{spwf})(0.0164)] \qquad (6.6.9)$$

where

$$C^* = \left(\frac{1}{C^n} - \frac{1}{C_c^n} \right)$$

c_R = cost savings by rehabilitation, $/ft.

c_R as calculated above must be compared with the cost of cleaning and lining, c_r. If the savings exceed the cost, then cleaning and lining is economical and should be implemented.

The parameter C^* is an indicator of the improvement in carrying capacity with cleaning and lining. Fig. 6.7 is included to assist the engineer in quickly determining C^* and to show that C^* is much more sensitive to C before cleaning and lining than the final C. This is fortunate, since the current value of C can be determined by testing but the final value can only be estimated.

EXAMPLE. Consider a 24-in. pipe which carries a peak flow of 9000 gpm and an average flow of 6000 gpm. Given the price of energy as 5 cents/kwhr, a wire-to-water efficiency of 70% (0.7), the cost to clean and line the main as $38/ft, and a series present worth factor of 9.82 (i.e., 20 years at 8%), is it economical to clean and line: (a) if no new pumping equipment is needed? (b) if new pumping equipment is needed? (c) if no new pumping is required but energy cost is 10 cents/kwhr? C is initially 70 but will be increased to 130 by cleaning and lining.

(a) C^* can be calculated as 2.63×10^{-4}. The savings in energy can be given by

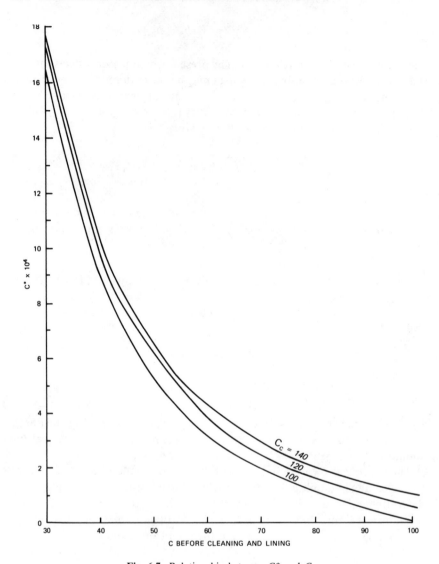

Fig. 6.7. Relationship between C^* and C.

$$c_e - c_{ec} = \frac{0.171(6000)^{2.85}(5)(9.82)(2.63)}{24^{4.86}10^4(0.7)}$$

$$= \$36.2/\text{ft}.$$

Since the cost to clean and line is \$38/ft, the project should not be undertaken.

(b) If new pumping equipment is required, it is necessary to determine the savings due to each foot of pipe lined. It can be given by

$$c_p - c_{pc} = \frac{6.76\,(2.63 \times 10^{-4})9000^{2.62}}{24^{4.87}}$$

$$= \$8.00/\text{ft}.$$

When this savings is added to the \$36.2/ft, the break even cost becomes \$44.2/ft and it becomes economical to clean and line.

(c) When the energy cost is increased to 10 cents/kwhr, it becomes economical to clean and line even if the price reaches \$72/ft.

The preceding example gives a good indication of the type of pipe that is a likely candidate for cleaning and lining. For example, in this case the velocity is 4.3 ft/sec and hydraulic gradient is 7.7 ft/1000 ft at average flow.

In the case of most looped systems, when a pipe is cleaned and lined it will carry flow that was originally carried by other pipes following parallel routes. In such cases both the cleaned pipes and the other pipes are served by the same pumping station. Thus the head that the pumping station must provide to all water is now decreased by the cleaning and lining. This means that there is a different Q_a in Eq. (6.6.4) before and after cleaning and lining and the savings in energy cost due to cleaning and lining are

$$c_e - c_{ec} = \frac{0.171\,P(\text{spwf})\,Q_t}{D^m e}\left(\frac{Q_a^n}{C} - \frac{Q_{ac}^n}{C_c}\right) \tag{6.6.10}$$

where

Q_{ac} = average flow carried by pipe after cleaning and lining, gpm

Q_t = flow carried by lined pipe and parallel pipes, gpm.

The problem in evaluating Eq. (6.6.10) is that for all but the simplest situations Q_{ac} must be determined by a network model and Q_t cannot be determined precisely. A similar situation exists for the pumping equipment savings cost which must be written

$$c_p - c_{pc} = \frac{6.76 Q_{tp}^{0.77}}{D^m}\left(\frac{Q_p^n}{C} - \frac{Q_{pc}^n}{C_c}\right) \tag{6.6.11}$$

where Q_{pc} = peak flow carried by the pipe after cleaning and lining, gpm.

The above equations appear fairly easy to use, but in practice there is no simple way to determine Q_t and Q_{tp}. Ignoring the effect of the parallel pipe will cause the cleaning and lining project to be undervalued, while accounting for the increased flow by using the final flow for Q_a in Eq. (6.6.10) will overestimate the savings. The following example for a simple case illustrates the size of the savings.

EXAMPLE. Consider two identical 30-in. parallel pipes, each with an initial C of 60, carrying a combined flow of 12,000 gpm. If, after cleaning, the C for one of the pipes is increased to 120, calculate the energy savings: (a) considering the lined pipe only ($Q_a = 8000$ gpm); (b) ignoring the parallel pipe effect ($Q_a = 6000$ gpm); and (c) accounting for both pipes. The series present worth factor is 9.8, the price of energy is 5 cents/kwhr, and no pumping equipment is required.

The initial flow distribution is 6000 gpm in each pipe. After cleaning and lining, it is 4000 and 8000 gpm, respectively. Recalling that

$$c_e - c_{ec} = \frac{(5)(9.8)(0.171)3.71 Q_a^{2.85}}{(0.6)(30^{4.87})(10^{-4})}$$

the answer to part (a) can be found using $Q_a = 6000$ gpm: \$19.4/ft. The answer to part (b) can be found by including the increase in flow ($Q_a = 8000$ gpm): \$43.9.

The answer to part (c) requires using a Q_t of 12000 gpm and the following formula:

$$c_e - c_{ec} = \frac{(5)(9.8)(0.171)(12000)}{(0.6)(30^{4.86})(10^{-4})} \times \left[\left(\frac{6000}{60} \right)^{1.85} - \left(\frac{8000}{120} \right)^{1.85} \right]$$

$$= \$28.5/\text{ft}.$$

The results of this simple example suggest a shortcut approach for calculating the savings for complex networks. That is, if a flow of 7000 gpm were used in parts (a) and (b), it would have given a savings of \$29.9/ft, which is a much better approximation to the correct answer than either of the other simplifications.

6.7.   COMPARISON WITH PARALLEL PIPE

If it is not possible to provide adequate head at the end of a pipe, then the pipe must be cleaned and lined, parallel pipe must be installed, or storage or booster pumping must be provided. In most cases cleaning and lining or parallel piping is the most economical alternative. To attain sufficient carrying capacity, it may not be necessary to clean and line or parallel the entire pipe, but rather just enough of the pipe to meet the target head loss. Upgrading only part of a pipe represents the least cost solution for either paralleling or cleaning and lining, so this should serve as the basis for selecting the method to be used. However, in many cases slightly more of the pipe may be upgraded than is really necessary because of access or terrain limitations, or simply to have a margin of safety.

In the following sections criteria for determining the length of pipe to be cleaned and lined are presented. The problem of parallel pipe optimization is somewhat more complex, since there are two design variables (diameter and length) which must be determined simultaneously to select the optimal strategy. A technique is developed below to select the diameter and length of the parallel pipe to arrive at a cost which can be compared to the cleaning and lining cost.

To determine the length of pipe to be cleaned and lined, knowing the design flow and head loss, it is necessary to write the energy equation to sum the head loss over the section to be cleaned and lined and the remainder of the pipe. The minimum cost results when the head loss constraint is exactly met:

$$h = \frac{K(L - L_c)}{D^m}\left(\frac{Q}{C}\right)^n + \frac{KL_c}{D^m}\left(\frac{Q}{C_c}\right)^n \qquad (6.7.1)$$

where

h = allowable head loss at design flow, ft
K = constant in head loss equation
L = total length of pipe, ft
L_c = length of pipe to be cleaned and lined, ft
D = diameter of pipe, in.
m = exponent on diameter (4.86 for Hazen–Williams equation)
Q = design flow, gpm
C = initial C factor
C_c = C factor after cleaning and lining
n = exponent on flow in head loss equation (1.85 for Hazen–Williams equation).

Solving Eq. (6.7.1) for the length to be cleaned and lined gives

$$L_c = L\left[\frac{\left(\dfrac{hD^mC^n}{KLQ^n}\right) - 1}{\left(\dfrac{C}{C_c}\right)^n - 1}\right]. \qquad (6.7.2)$$

If L_c turns out to be negative, cleaning and lining is not required, except possibly to reduce energy costs as described in the previous section. If $L_c > L$, the head loss criterion cannot be reached by cleaning and a parallel pipe is required. Dividing through by L and recognizing the dimensionless factor Y reduces Eq. (6.7.2) to

$$L_r = \frac{1 - Y}{1 - C_r^{-n}} \tag{6.7.3}$$

where

$\quad L_r$ = fraction of pipe to be cleaned and lined
$\quad Y$ = dimensionless head loss, $hC^n D^m / LQ^n K$
$\quad C_r$ = ratio of C after to C before cleaning and lining (C_c/C).

EXAMPLE. Given 5000 ft of 36-in. pipe with a C of 80 along which there must be less than 30 ft of head loss, find the length of pipe that must be cleaned and lined to meet this goal for a design flow of 20000 gpm, and $C = 140$ after cleaning and lining.
  First calculate Y to determine if cleaning and lining is required:

$$Y = \frac{30}{5000} \left(\frac{80}{20000}\right)^{1.85} \frac{36^{4.86}}{10.4}$$

$$= 0.80.$$

Since $Y = 0.80$, the head loss criterion cannot be met without cleaning and lining or some other measure. Next, calculate the relative improvement in C resulting from cleaning and lining and use that value to find the fraction of pipe to be cleaned and lined:

$$C_r = (140/80) = 1.75$$

$$L_r = \frac{(1 - 0.80)}{(1 - 1.75^{-1.85})} = 0.31.$$

Therefore the length of pipe to be cleaned and lined is

$$L_c = 5000(0.31) = 1550 \text{ ft.}$$

  The cost of the cleaning and lining alternative is simply the length to be cleaned and lined times the unit cost for the work.
  The cost of the parallel piping alternative depends on the unit price of pipe and the diameter and length of that pipe. The following equation, similar to Eq. (6.7.1), must hold for the parallel pipe (i.e., the sum of the head losses in the parallel section and the section not paralleled must be equal to the allowable head loss for least cost):

$$h = \frac{KL_p}{D_p^m} \left(\frac{Q_p}{C_p}\right)^n + \frac{K(L - L_p)}{D^m} \left(\frac{Q}{C}\right)^n \tag{6.7.4}$$

where

$$Q_p = \text{flow in the parallel pipe, gpm}$$
$$Q = \text{total flow in pipe, gpm}$$
$$L_p = \text{length of pipe paralleled, ft}$$
$$D_p = \text{diameter of parallel pipe, in.}$$
$$C_p = C \text{ for parallel pipe.}$$

The head loss in the parallel pipe and in the pipe it is paralleling must be equal. It is possible to use this fact to solve for the flow in the parallel pipe, which is not known, in terms of other parameters which are known, or can be determined; that is,

$$Q_p = \frac{Q}{1 + (C/C_p)(D/D_p)^{m/n}}. \tag{6.7.5}$$

Substituting this equation back into Eq. (6.7.4), nondimensionalizing as was done in the previous case, and solving for the ratio of the diameter of the parallel pipe to the diameter of the original pipe, yields

$$D_r = \left\{ \left[\left(\frac{L_r}{Y + L_r - 1} \right)^{1/n} - 1 \right] \frac{1}{C_r} \right\}^{n/m} \tag{6.7.6}$$

where

$$D_r = D_p/D$$
$$L_r = L_p/L$$
$$C_r = C_p/C.$$

The problem at this point is that there are many combinations of D_r and L_r that will satisfy Eq. (6.7.6). It is necessary to use cost of the parallel pipe to select the best alternative to compare with cleaning and lining.

The key to solving this problem manually is to take advantage of the facts that pipe can be purchased only in commercially available discrete diameters, so the optimal combination of D_r and L_r will usually occur where $L_r = 1$. Therefore solving the problem is simply a matter of finding the D_r for $L_r = 1$, i.e., paralleling the entire pipe, rounding that number to the next larger commercially available size, and calculating the length as shown below:

$$L_r = \frac{1 - Y}{1 - \left(\dfrac{1}{D_r^{m/n} C_r + 1} \right)^n}. \tag{6.7.7}$$

Then calculate the cost for this size and length of pipe. Next, reduce the pipe size to the next smaller diameter and recalculate the length to be paralleled and the associated cost. Continue this procedure until the costs begin to increase. Select the pipe size that resulted in the lowest cost. This procedure is summarized in Fig. 6.8. The parameter z in the figure refers to the cost of the parallel pipe for pipe of that diameter, and can be calculated as

$$z = aLL_r(DD_r)^b \tag{6.7.8}$$

where

$$z = \text{cost of parallel pipe, \$}$$
$$a,b = \text{regression constants in cost function}$$

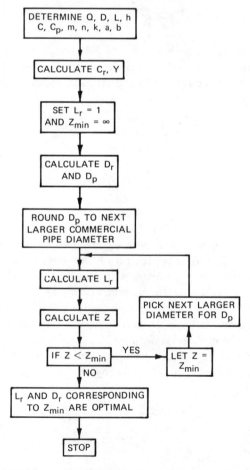

Fig. 6.8. Procedure for selecting optimal parallel pipe.

or

$$z = AL_r(D_r)^b \qquad (6.7.9)$$

where $A = aLD^b$.

EXAMPLE. Consider an existing 6000-ft-long, 12-in.-diameter pipe ($C = 60$) which must carry 2000 gpm with a head loss of 10 ft. Calculate the optimal length and diameter of a parallel pipe which will meet this goal. Use $C = 130$ for the parallel pipe, $a = 1.0$, $b = 1.4$, $K = 10.4$, $m = 4.87$, and $n = 1.85$.

1. Calculate C_r and Y:

$$C_r = 130/60 = 2.167$$

$$Y = \frac{10}{1000}\left(\frac{60}{2000}\right)^{1.85}\frac{12^{4.87}}{10.4} = 0.264.$$

2. For $L_r = 1$, calculate D_r:

$$D_r = \left\{\left[\left(\frac{1}{0.264}\right)^{0.54} - 1\right]\frac{1}{2.167}\right\}^{0.38} = 0.76$$

$$D_p = D_r D = 0.76(12) = 9.12 \text{ in.}$$

3. Round D_p to 10 in. Then

$$D_r = 10/12 = 0.833.$$

4. Calculate L_r:

$$L_r = \frac{1. - 0.264}{1 - \left(\dfrac{1}{0.83^{2.63}2.167 + 1}\right)^{1.85}} = 0.928.$$

5. Calculate z:

$$z = 1.0(1000)(0.928)(10)^{1.4} = 23210$$

This becomes z_{min}.

6. The next candidate for D_p is 12 in. (i.e., $D_r = 1$):

$$L_r = \frac{1. - 0.264}{1 - \left(\dfrac{1}{1^{2.63}2.167 + 1}\right)^{1.85}} = 0.835.$$

7. Calculate z for $D_p = 12$ in., i.e.,

$$z = 1.0(1000)(0.835)(12)^{1.4} = 27073.$$

8. Since this cost is larger than the cost for the 10-in. pipe, the 928-ft 10-in. line is the optimal parallel line and the solution is complete. If the 12-in. pipe was less expensive, it would be necessary to calculate the cost of a 14-in. parallel pipe to determine if additional savings could be realized by increasing the diameter even more.

In general, the parallel pipe that runs the entire length of the line being paralleled will be the optimal size for smaller values of Y (i.e., $Y < 0.5$). Only when the shortfall of carrying capacity in the original main is marginal is it economical to parallel part of the pipe. In those cases, however, it is usually more economical to clean and line the pipe.

Since in virtually all cases in which a parallel pipe is economical the entire pipe will need to be paralleled (i.e., $L \approx 1$), the optimal pipe diameter can be approximated by setting $L_r = 1$ in Eq. (6.7.6) to give

$$D_r = \left\{ \left[\left(\frac{1}{Y} \right)^{1/n} - 1 \right] / C_r \right\}^{n/m}. \tag{6.7.10}$$

Fig. 6.9 is a graphical solution to Eq. (6.7.9) for $1/n = 0.54$ and $n/m = 0.38$. To select a reasonable value for D_r, calculate Y, and C_r, and look up D_r on Fig. 6.9. Multiply D_r by D to obtain D_p which should be rounded up to the next commercially available pipe size to determine the size of the parallel pipe.

Once the cost of the parallel pipe is determined it must be compared with the cost of cleaning and lining to decide which approach to restoring carrying capacity is more economical. In general smaller pipes should be paralleled while larger ones should be cleaned and lined. This is due to the greater economy of scale with respect to diameter for cleaning and lining as opposed to buying new pipe.

Figure 6.10 gives an overview of the comparison between paralleling and cleaning and lining based on $c_r = \$28/\text{ft}$, $m = 4.87$, $n = 1.85$, $k = 10.4$, $a = 1.0$, and $b = 1.4$, assuming that the C of a new pipe and a cleaned and lined pipe are the same. This figure merely shows the impact of the different variables on the decision. The actual decision for a particular utility will be based on costs for the site specific conditions encountered. For example, laying new pipe in a congested downtown area will be significantly more costly than laying new pipe in a new residential area, while the costs of cleaning and lining such a pipe will only be slightly greater in a downtown area since the access holes have a small impact on traffic.

In addition to showing that paralleling is more economical for smaller pipes, Fig. 6.10 shows that there is an upper limit to how smooth a pipe can be

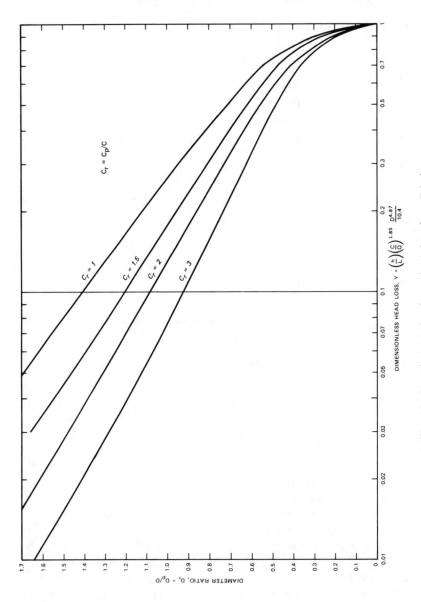

Fig. 6.9. Graphical solution for optimal size of parallel pipe.

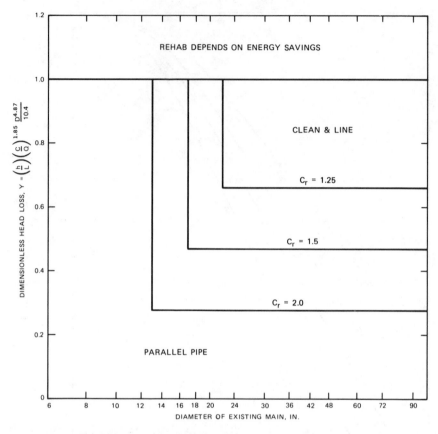

Fig. 6.10. Overview of decisions for restoring carrying capacity.

made by cleaning and lining. There are some cases (i.e., below the horizontal lines in Fig. 6.10) for which cleaning and lining simply will not be able to meet target head loss. This indicates that a pipe must be fairly rough to begin with for cleaning and lining to make such improvement.

Since Y is also affected by the design flow rate, cleaning and lining will be more attractive in areas with little expected growth in water use, while paralleling will be more attractive in lines supporting fast-growing areas.

In applying the procedures described above to a real system, it is necessary to carefully select which mains are to be considered for upgrading. The mains selected should radiate out from the treatment plant/pumping station, as opposed to pipes perpendicular to that direction which are used to close the grid. Attention should be focused on pipes with large hydraulic gradients, and/or pipes which must support growth in the system.

The results of this analysis are highly dependent on Y, which in turn is very sensitive to the desired head loss h (or hydraulic gradient h/L) and design flow Q. Both values should be selected based on future head and flow requirements rather than current values, and should take into account the overall expansion plans for the system. If only a portion of a line is to be cleaned and lined or paralleled, that portion should be on the upstream end of the line because benefits accrue only to parts of the system downstream of the improvements. However, in many cases the section may need to be selected based on ease of access. Parallel pipe need not be laid in the same street as the line it is paralleling, but can be laid along some other parallel route where the new pipe can do the most good.

6.8   COMPARISON OF LINING VS. CLEANING ONLY

Lining a pipe will maintain the increased carrying capacity indefinitely. In cases, however, in which the water quality from a corrosion or scaling standpoint is good, it may not be economical to line the pipe after it has been cleaned. Carrying capacity in the unlined pipe will eventually deteriorate to the point where there is no benefit from the cleaning, and the pipe will need to be cleaned again (even though it may not be). The question is: When is lining economical?

An economic evaluation of lining should be based on a comparison of cleaning and lining vs. the present worth of cleaning to achieve the same benefits. A pipe should be lines if:

$$c_L < \sum_{j=1}^{N} c_c/(1 + i)^{(j-1)t} \tag{6.8.1}$$

where

$\quad c_L$ = cost of cleaning and lining, $/ft
$\quad c_c$ = cost of cleaning only, $/ft
$\quad i$ = interest rate
$\quad N$ = number of years in planning horizon, yr
$\quad t$ = time until benefits of cleaning project are lost, yr

The key to this evaluation is determining the time t when the carrying capacity of the unlined pipe is substantially reduced. With fairly corrosive water, this can be as short as 5 years. This does not mean that the C-factor will return to its pre-cleaning value but that the pipe carrying capacity is degraded. t will be much lower than the age of the pipe when cleaning is initially required. Presently this is essentially a judgment decision.

Table 6.1   Critical Values of Ratio of Cleaning and Lining to Cleaning Cost (P).

Interest Rate (%)	4	8	t 12	16	20
2	10.4	6.5	4.7	3.7	3.0
4	6.6	3.7	2.7	2.1	1.8
6	4.8	2.7	2.0	1.6	1.4
8	3.8	2.2	1.7	1.4	1.3
10	3.1	1.9	1.5	1.3	1.2
12	2.7	1.7	1.3	1.2	1.1
14	2.4	1.5	1.3	1.1	1.1
16	2.2	1.4	1.2	1.1	1.05

Equation (6.8.1) can be rearranged to give

$$c_L/c_c < P = \sum_{j=1}^{N} 1/(1 + i)^{(j-1)t} \qquad (6.8.2)$$

P is the critical value of the ratio of cleaning and lining cost at which it becomes economical to line the pipe. P is a function of interest rate and t, and the values are given in Table 6.1. If the interest rate is 10 percent and the time at which carry capacity is lost is 8 years, then Table 6.1 indicates that P is 1.9. Therefore, if the cost of cleaning and lining is more than 1.9 times the cost of cleaning alone, then lining is not economically justifiable.

The evaluation described above should only be considered as a rough guide when deciding about pipe rehabilitation. In addition to the uncertainty in determining t, there are other considerations including: 1. cleaning and lining contracts usually call for provision of bypass piping during the job; 2. lining can seal off minor leaks; 3. cleaning without lining only temporarily solves red water problems; and 4. water quality may have changed since the original scaling or corrosion occurred. The evaluation described above should give the engineer an appreciation of the tradeoffs involved in the decision.

REVIEW QUESTIONS

1. Why do pipes lose their carrying capacity? Why is loss of carrying capacity more likely for unlined metal pipes?
2. Why is it beneficial to line a pipe after it has been cleaned?
3. Why must butterfly valves and check valves be removed before a pipe can be cleaned?
4. In what cases can new elevated storage be used to overcome the effects of lost carrying capacity?

5. Why does cleaning and lining involve less disruption of traffic than installation of new pipe?
6. How does the situation where $Y > 1$ differ from the situation $Y < 1$ insofar as techniques for restoring carrying capacity are concerned?
7. How does the overall size of the job affect the cost of cleaning and lining? How does the number of services affect the cost?
8. The price of energy used in calculating future energy costs should be the price in what year?
9. Does a value of 0.001 for C^* indicate that the carrying capacity will be greatly or slightly improved by cleaning and lining?
10. Why is the average flow rather than peak flow used as a basis for energy costs in the formulas for calculating energy savings due to cleaning and lining?
11. In the case of a pipe that is part of a looped system, why is it not correct to use the same flow in the pipe both before and after cleaning and lining?
12. Why is selecting the length of pipe to be paralleled more involved than selecting the length of the pipe to be cleaned and lined?
13. Why is paralleling more economical than cleaning and lining for small pipes?
14. Why is there a lower limit on the Y values for which cleaning and lining is feasible?

PROBLEMS

1. Determine whether to clean and line a 2000-ft-long, 12-in.-diameter main designed to carry 1500 gpm, which usually carries only 800 gpm. The existing C factor is 80 but cleaning and lining would raise it to 125. Additional pumping capacity will be

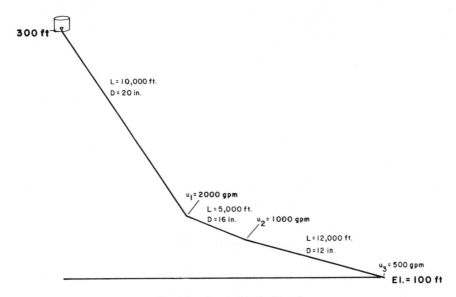

Fig. 6.1p. System for Problem 2.

required in 4 years and the allowable head loss in the line is 35 ft. The cost of power during the design period averages 10 cents/kwhr and the cost to rehabilitate is $25/ft. Use spwf = 8.5 (10% for 20 years) and pwf = 0.683 (10% for 4 years) and a wire-to-water efficiency of 70%.

2. Given the system shown in Fig. 6.1p, find Y for each section of pipe for: (1) the case where the peak use as shown in the figure occurs and 40 psi must be maintained, and (2) a fire demand of 1500 gpm must be met at the end of the pipe and 20 psi must be maintained. Start calculating Y from the upstream end of the system for each case. If the head at the beginning of a pipe is calculated to be lower than the required head, assume the head at the node will be increased to meet the head required. Therefore use the minimum head as the upstream head in those cases instead of the actual head.

3. Consider a 24-in., 3-mile-long pipe with existing $C = 70$. The beginning of the pipe is at a reservoir with low-water elevation of 400 ft, and the end is a community at an elevation of 145 ft which desires 80 psi. Calculate the head loss and Y and compare the cost of cleaning and lining at $35/ft with the cost of installing 14-, 16-, and 18-in. parallel pipe at a cost of $45, $50, and $60/ft, respectively. The new pipe and the cleaned and lined pipe will have $C = 140$. The pipe must carry 10,000 gpm peak flow, 5000 gpm average flow, and the cost of energy is 8 cents/kwhr.

REFERENCES

Merril, D.T., R.L. Sanks, and C. Spring, 1978, *Corrosion Control by Deposition of Calcium Carbonate Films,* American Water Works Association, Denver, Co.

Morgan, J.M., T.M. Walski, and M.W. Corey, 1984, *Simplified Procedure for Calculating Chemical Dose for Water Conditioning,* WES Technical Report, Vicksburg, Miss.

7 | PIPE BREAKS AND WATER LOSS

7.1. INTRODUCTION

Water under pressure in a distribution system wants to get out of the pipe. It is the utility's responsibility to insure that this only occurs at points of authorized water use. Authorized use includes water removed from the system for use by paying customers, authorized nonpaying customers, and municipal activities such as fire fighting and street washing. The difference between water produced and authorized use results from leakage and unauthorized use. This water is considered lost.

Another important term is unaccounted-for water, which may be defined as the difference between water produced and water used at metered services. Unaccounted-for water includes lost water, water used at unmetered services, and net underregistration at meters. It is important to remember the distinctions between unaccounted-for water, lost water, and leakage.

Water leaks from the distribution system through openings which range in size from ruptures of large mains to drips from loose nuts in meter boxes. In this chapter, main breaks will refer to actual breaks in the pipe wall in major pipes, as distinct from joint leaks and service line leaks. Main breaks and joint leaks require costly excavation and clamps or patches to repair leaks, while leaks at hydrants and meter boxes can often be repaired by tightening or replacing fittings.

There are two fundamental questions which must be answered by every utility: (1) Should a leak detection program be implemented? (2) Should main breaks continue to be repaired or should some other remedial measure (e.g., cathodic protection or replacement of sections of mains) be implemented? These questions plus many associated considerations are addressed in this chapter.

First, the types of breaks and their causes are described, then leak detection methods are presented. This is followed by a discussion of remedial measures. The costs of detection, repair, and other associated impacts are considered and methods for economic evaluation of the solutions to problems of leakage are developed. Finally, the subject of evaluating unaccounted-for water is addressed.

7.2. CAUSES OF BREAKS AND LEAKS

No program to remedy problems of pipe breaks and leaks should be instituted before the principal causes of the breaks that are occurring are determined. Unfortunately, this is complicated by the fact that there are often several causes contributing to a given break. For example, a pipe may fail because of excessive load, but the location of the break may be due to a clay ball in the bedding. Similarly, waterhammer may trigger a joint leak in a pipe that was not adequately restrained during installation.

Knowing the causes of past breaks and leaks is important in deciding on strategies for responding to breaks. Corrosion-induced breaks will require cathodic protection or replacement with protected pipe, or changes in water treatment practices, while pipe that was damaged in shipment may simply need to be replaced. Each of the causes of breaks is described in the following paragraphs.

A high break and leak rate immediately after a pipe is installed indicates that: (1) there were defects in manufacture, (2) the pipe was damaged in shipment, or (3) the pipe was laid improperly. In such cases the engineer must carefully examine every step of the installation procedure and have some of the pipe tested to determine where responsibility should be placed.

Improper tapping, such as not using a tapping saddle on a thin-walled pipe where one is required, can result in breaks radiating out from the tap. Most manufacturers specify the minimum wall thickness required for tapping without saddles. Failure to observe these specifications can lead to breaks.

Poor backfill and bedding can lead to breaks because loads on the pipe are not evenly distributed as was assumed during design. Rocks and hard mud balls in the soil bedding are usually the cause of these problems. When a break of this type is found, the bedding around the break should be examined to determine if this is an isolated case or if the bedding is generally substandard.

Even if the bedding was adequate when the pipe was installed, it may have been eroded away by a leak at a joint. When the bedding is washed away the pipe is no longer supported adequately and is more likely to break under external loads.

The extreme case of disturbed bedding occurs when the bedding is replaced by a structure or another conduit. This is especially common in congested urban areas where water mains may come in contact with sewers, subway structures, gas mains, electrical and phone conduits, and foundations for buildings and bridges. Repair of this type of break involves relocating the water main in a trench by itself or in a properly designed utility trench.

New construction in the vicinity of a pipe can alter the loads that a pipe must bear. Sometimes pipes are actually broken by the construction equipment; sometimes foundations are laid on top of a pipe or new roads are constructed (or old ones repaired) in such a way that a pipe must carry loads it was never designed to carry. Breaks resulting from this cause generally show up shortly after construction is complete.

Loads on pipes can increase from natural as well as man-made causes. Large depths of frost penetration can break pipes which were designed to handle normal loads. Expansive soils also impose severe demands on pipe strength. One can also count on pipe breaks near landslides or subsidence events. In some cases, broken water mains have even been the cause of subsidence.

In places where adequate protection was not provided, the water in the main itself can freeze and damage pipes and appurtenances. This usually can be prevented by eliminating dead ends so that water always keeps moving, or, in extreme conditions, by burying pipes deeper. Metal pipes can be thawed electrically, although it is important to be certain the current can flow between pipe sections. In some cases conductivity straps are required.

Waterhammer is sometimes sited as a cause of breaks, but this will not generally be a problem unless the pipe is weakened considerably by corrosion. Waterhammer is more likely to cause problems at joints near bends where the pipe is not adequately restrained. This type of leak is likely to show up after a fire or some other event where waterhammer is likely to occur. A rash of joint leaks in an area means that better thrust blocking or additional restraints such as tie rods or joint clamps are required.

Corrosion is the root, if not the immediate, cause of most breaks in metal pipes. Metals tend to want to return to their ore state. When the soil is dry and has a high resistivity, this process is inhibited. However, in wet soils, with low redox potential and low resistivity, corrosion will proceed rapidly. Corrosion pulls iron out of the pipes, leaving carbon behind, in a process known as *graphitization*. The pipe may look virtually the same, but it will not be nearly as strong.

While corrosion is a natural process, it can be accelerated by stray currents caused by grounding electrical devices to water pipes or impressed current cathodic protection of nearby gas mains or other structures. Fairly small stray dc currents can severely corrode a pipe after only a few years. Contact with metals which are less active can set up galvanic action which results in fairly rapid corrosion.

The utility should attempt to identify the cause of every break at the time it occurs. The answers to a few very simple questions can assist the engineer in the detective work of identifying the cause of breaks. These include the following: Are there seasonal trends indicating frost penetration or expansive soil as the culprit? Are nearby gas mains cathodically protected with impressed currents, or is there electrical equipment grounded to (or near) the mains? Do joint leaks occur immediately after fires or hydrant tests? Has there been new construction in the area? Are breaks increasing with time? The types of questions that should be answered for each break are discussed in more detail in Section 7.9.

The form of the break also gives clues to its cause. Small holes are usually due to corrosion. Circumferential breaks usually indicate excessive beam loading due to contact with structures or poor bedding. Joint leaks are generally due to motion caused by inadequate restraint, expansive soil, or subsidence.

7.3. LEAK DETECTION

Detecting leaks in a water distribution system involves a great deal of patience, skill, determination, and a bit of luck. It is analogous to finding the proverbial needle in the haystack, but the benefits from finding only a handful of leaks can easily outweigh the costs.

Leak detection methods can be grouped into several categories: (1) listening; (2) listening plus flow measuring, which is referred to as a *water audit;* and (3) hydrostatic testing. A fourth method, still commonly used today, is to wait until the water bubbles or bursts through to the surface. Reliance on this method alone results in missing many leaks, some of which may be fairly substantial, as the water may be able to find its way to a stream or sewer undetected.

The concept of a water audit follows the simple principle that the water produced for the system should equal the water accounted for, plus an allowance for unmetered authorized uses. A water audit of the entire system indicates how much water is probably leaking from the system. Leak detection requires that the exact location be determined. This cannot be done with a water audit alone, but application of water audit techniques to small portions of a system can isolate areas which have the most leaks, thus enabling listening devices to be used most effectively.

In conducting a water audit of a portion of a distribution system (referred to as a district), it is necessary to isolate that district from the remainder of the system so that water can only enter (or leave) that district through one or two mains which are metered.

In most systems, there is little or no metering of large mains, so it is necessary to install a meter for the water audit. Most large permanent meters are very expensive, so pitot tubes equipped with recording devices are generally used to measure flow. (Pitot tubes are described in greater detail in the next chapter.)

Another problem with water audits is the fact that it is often difficult to isolate portions of systems because of malfunctioning valves. This actually represents an indirect benefit of a water audit in that conducting the audit forces the utility to locate and repair defective valves. This saves the utility the problems of finding that the valves are defective when they are needed for an emergency shutdown. In addition to giving valves much needed exercise, a water audit provides an opportunity for utility personnel to observe how their system operates under a partial shutdown.

The primary purpose of a water audit is not to test valves but to measure flows under controlled conditions. This should be done for at least a 24-hour period. According to Cole (1980), the best indicator of leakage is the *minimum night ratio*, which is defined as the ratio of the lowest water use rate at night to the average daily use. Note that average use in the district under consideration is the important parameter, not flow through the district. For districts with no storage and no major wet industry operating at night, the minimum night ratio can be given by

$$MNR = \frac{Q_{min}}{Q_{av}} \qquad (7.3.1)$$

where

$$MNR = \text{minimum night ratio}$$
$$Q_{min} = \text{minimum inflow to district at night, gpm}$$
$$Q_{av} = \text{average daily inflow to district, gpm.}$$

When there are storage tanks or major wet industries operating during the night, the industrial use should be subtracted from the total use as follows

$$MNR = \frac{Q_{min} - I_{night} - S_{fill}}{Q_{av} - I_{av} - (dS/1440)} \qquad (7.3.2)$$

where

$$I_{night} = \text{night industrial use, gpm}$$
$$I_{av} = \text{average industrial use, gpm}$$
$$S_{fill} = \text{rate of filling of tanks in district, gpm}$$
$$dS = \text{change in storage over day, gal.}$$

This means that meters at large industries must be read as part of the water audit.

Since domestic and commercial water use is generally very low at night, one would expect water consumption in general to be low at night. However leakage remains constant over the course of the day, and may even be higher at night because of generally higher pressures in the system. So if the minimum nighttime ratio is greater than roughly 50%, there is reason to suspect considerable leakage in the district. If it is below 35%, there is probably very little leakage.

Siedler (1984) suggested that it was possible to distinguish between meter underregistration and leakage as a source of unaccounted-for water by first dividing metered domestic water use by metered population. If that value is less than 50 gal/capita/day then meter underregistration is likely to be significant. Next subtract metered industrial, commercial and public use from total production and divide the remainder by the population served. If the ratio is greater than 70 gal/capita/day then leakage is likely to be significant. Some systems can have leakage and underregistration problems.

Whether a water audit is used or not the final steps in leak detection involve listening for leaks. When water escapes from an orifice, it causes a vibration in the 500–800 Hz (Hertz) range (Heim, 1979). This sound travels along the pipe wall and can be heard a considerable distance away by an observer with the proper equipment. This sound is the key to identifying the presence of a leak.

Other sounds (in the 20–250 Hz range) are caused by water striking the soil and swirling around in the cavity it creates. This sound does not travel well along the pipe, and is therefore useful in pinpointing the leak.

The earliest leak detection devices were nothing more than a stick or rod which could carry sound from the pipe or appurtenance to the ear of the listener. In time, telephone earpiece devices and stethoscope devices were added to improve sound detection. At present electronic amplification and filtering techniques have been added to the mechanical devices to give better results with less background noise. A typical sonic leak detection device is shown in Fig. 7.1. New correlation devices have been developed which use sound intensities from two locations to help pinpoint leaks.

Generally, listening is done at any point that can be reached by the device, including valves, meters, and hydrants. When a leak is suspected along a pipe, it is also possible to use geophones similar to those shown in Fig. 7.2 to listen to the pavement itself to pinpoint the leak. In some cases, however, it is necessary to drive a rod to the top of the pipe to confirm a leak.

When leaks are not suspected in an area, it may be adequate to skim the area by listening only to random hydrants. In a thorough leak detection survey, it is necessary to listen at every possible listening point. Crews generally consist of two persons, one operating the listening device while the other assists with equipment and traffic control.

Usually a crew will walk one side of the street from their truck and return

Fig. 7.1. Sonic leak detection device. [Courtesy Heath Consultants.]

down the other side of the street. With some of the more exotic devices, the equipment must be housed in a van which accompanies the crew.

Leak detection by listening is qualitative, as there is virtually no correlation between size of leak and intensity of sound. The sound is influenced by such factors as pipe material (metal pipe conducts sound better), soil type, and leak configuration.

When a leak is found on a main, a crew is called to make the repair. Leaks attributed to improperly closed hydrants and loose nuts at meters can be corrected by the listening crew. If a leak is found on a service line, it is possible to shut off the service at the meter and lower the pressure on the customer's side of the line by opening a faucet. Once the pressure drops the faucet is closed, and if the leak is on the customer's side of the line, the sound will cease. If the leak sound continues, then the leak is on the utility's side of the meter. In general, leaks will not make an appreciable sound if the line pressure is less than 15 psi.

There should be a method for repair crews to provide leak detection crews with feedback as to the size and type of leak discovered. This will assist the leak detection crew in honing their skills and will reduce the number of dry holes.

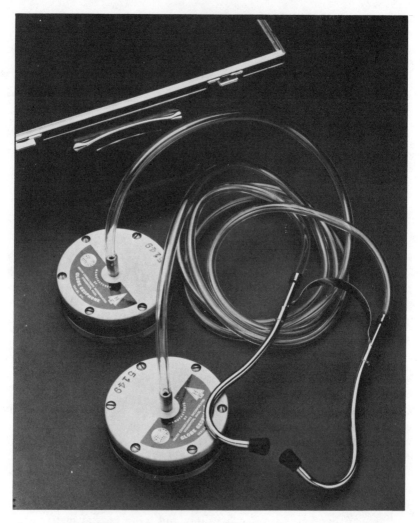

Fig. 7.2. Geophones for leak detection. [Courtesy Heath Consultants.]

Listening work can be done on contract with a firm specializing in such testing, with utility personnel using rented equipment or with utility personnel with equipment owned by the utility. Laverty (1979) reported that East Bay Municipal Utility District personnel became experts in using such equipment after little training. In a study for the State of California, Boyle Engineering (1982) tested several types of listening devices, all of which were successful in detecting leaks.

Hydrostatic testing, similar to that performed when new pipe is installed, can also be used to test existing water systems. Kocol (1972) and McPherson

(1983) described use of these tests in Milwaukee and Rochester, and reported that the tests enabled them to find weak sections of pipe and leaks by subjecting the pipe to high pressures (300 psi). This type of testing can only be done in areas with few customers, as all service connections must be closed before the tests to prevent damage to customer plumbing. Causing mains to fail when repair crews are on hand is highly preferred to having the mains fail at 3 am on a Saturday night in February.

One final technique which is used to locate leakage is to look into sewers for large flows of clear water. This can be helpful in detecting large leaks which enter sewers. Such water may, however, also be an indication of infiltration of groundwater. The presence of fluoride in the water is a good indication that the source of the water is the distribution system.

7.4. EVALUATION OF LEAK DETECTION

A leak detection program involves a relatively small cost when compared with the total cost of operating of a utility and usually saves more than it costs. The decision to institute a leak detection program or conduct a leak detection study should be based on a realistic estimate of the benefits and costs of such work. Utilities with low leakage rates and little need to construct new facilities will benefit less from such a program than those with high leakage and the immediate need to construct new, costly facilities. A cost-benefit analysis should be conducted before a leak detection study to be certain the money is spent wisely. The individual cost items and methods for calculating expected benefits, discussed in the following paragraphs, are based primarily on papers by Boyle Engineering (1982), Kingston (1979), and Moyer et al. (1983).

The cost of a leak detection survey depends on whether it is done by utility or contractor personnel, whether equipment is bought or rented, whether a water audit is part of the survey, and whether the survey is a one-time study or an ongoing program.

The following items should be considered in developing a cost estimate:

1. Utility labor
2. Utility trucks and communications
3. Listening equipment
4. Water audit equipment
5. Water audit setup
6. Water audit labor
7. Repair of leaks
8. Contract labor
9. Contract equipment use.

Regardless of whether contractors or utility personnel are used to do the actual listening, there will always be a considerable investment of utility man-hours for procuring contractors and equipment, acting as guides, performing the acutal listening and water audit, operating valves, repairing leaks, and working with customers to minimize inconvenience. These costs are best estimated by multiplying crew size by number of hours, after separating the audit, listening, and repair activities into separate categories. The cost of trucks and compressors and earth moving equipment for detection and repair can be estimated by multiplying the number of hours by the unit cost of the equipment.

Estimating the number and size of leaks which will be found involves considerable judgment. A good starting point is to estimate the amount of water loss attributed to leakage which will be found during the survey. For example, if leakage is estimated at 100,000 gpd, a survey can usually locate about half of this leakage (50,000 gpd). About half of this volume will result from breaks and joint leaks having an average leakage rate of 10,000 gpd (7 gpm). Most will be smaller, but one or two large leaks can significantly increase the average. Repair of the large leaks requires a significant repair effort. (For this example, there should be about $50,000/10,000 = 5$ of these leaks.) The other half of the leakage generally results from smaller leaks around meters, valves, and hydrants. A typical average leakage rate for these small leaks is about 1,000 gpd. (For the example, there should be about 50 of these leaks.) The labor involved with repairing this kind of leak is small, since it involves no excavation and can be done by the listening crew in many cases. Once the number of leaks is estimated, the cost can be estimated from repair crew size and cost.

If the survey is going to be a one-time operation, it may be economical to rent rather than purchase the listening equipment. Certain devices can only be used under contract with the equipment owners. Weekly rental rates tend to be on the order of 10% of the purchase cost, so purchase is often the most economical alternative if the survey is going to last for more than 10 weeks. The break-even point for each piece of equipment will vary, but in general it is worthwhile to purchase this equipment, as it may come in handy after the survey.

A water audit involves isolating districts within the system by closing valves. The cost of a water audit can be estimated at roughly three man-days per district for closing and later opening the valves, plus the cost of contractor equipment to conduct the flow measurements and analyze data. If this work is to be done often, it may be worthwhile for the utility to purchase the pitot tube and differential pressure recorders and train personnel to use them. The largest cost item for a water audit is establishment of flow measuring stations. Occasionally there is a vault or manhole on the line to be tested so that the only cost is for installation of a tap. However, in most cases, considerable effort in excavation and shoring is required to establish a flow measuring station. If water audits are to be conducted in the future, it is wise to make the metering station permanent by providing a manhole over it. This will greatly reduce the cost of future water audits.

The various costs should be combined to give the cost of the total survey for a one-time effort or the annual cost of a continuing program. The actual cost for repairs will decrease with time, since fewer breaks will typically be found each year.

The benefits of leak detection and repair depend on whether the utility purchases treated water at adequate pressure or supplies its own water. For the purchased water case, benefits can be estimated as the product of the quantity of water saved and the unit price. If the utility must obtain its own water, or treat or pump purchased water, the price is no longer a good indicator of the benefits, since it reflects many fixed costs, such as debt retirement and billing, which are not reduced by leak repair. If the utility is not expanding, the only savings possible are in variable operation and maintenance costs, such as pumping energy and chemicals, which economists refer to as short-run costs. If, however, the utility is planning to build new facilities or increase capacity of existing facilities and leak repair enables the utility to downsize or delay construction of these facilities, very large benefits can be realized in what economists call long-run costs. To realize these long-run cost savings, the leak repair program must actually impact system capacity and not simply lower the quantity of water to be pumped and treated.

Long-run cost savings can be estimated by comparing the difference in the cost of constructing and operating new, or expanded facilities with and without the leak detection and repair program. For example, a utility with an 18 mgd capacity will need to increase capacity by 7 mgd to 25 mgd at a cost of $15 million if leakage is not reduced. When the flow is 25 mgd, it is estimated that leakage is 5 mgd, and a leak detection program can reduce this leakage to 2 mgd. This program can reduce the size of system expansion from 7 mgd (25 − 18) to 4 mgd and reduce the cost by $7 million. This $7 million is the long-run benefit of the program. Of course, to realize the full benefit, leak detection must be an ongoing activity. From the above example, leak detection can be seen to offer the greatest advantage to utilities with high loss rates and the need to expand. Utilities not experiencing growth will only realize short-run cost savings and savings in purchased water. In estimating long-run benefits, the quantity of water saved should be subtracted from system capacity, while in the case of short-run savings and purchased water, the reduction in leakage should be subtracted from actual water purchased or pumped. There should be a cutoff time after which benefits no longer accrue, since the leak would have been found by other methods. This cutoff time can be estimated as one year for large leaks and up to five years for small leaks.

The water loss benefits from leak detection can be estimated from the following formula

$$B = c^*[1 - (1 - \Delta Q/Q^*)^b] + 525\Delta QPt \qquad (7.4.1)$$

where

B = benefits from water saved by leak repair, $

c^* = present worth of future construction and O&M on new facilities, $

ΔQ = reduction in leakage, gpm

Q^* = additional capacity added to the system, gpm

b = economy of scale factor (usually 0.7 for water facilities)

P = price of water purchased by utility plus unit cost of pumping energy and chemicals, $/1000 gal

t = difference between the time when a leak is found with detection program and when it would be found otherwise, years.

The preceding formula is based on the assumption that new facilities will be downsized because of water conservation resulting from the leak detection and repair effort. It is, however, also possible to eliminate the need for facilities, or to delay construction significantly. Methods to calculate these costs savings are given by Walski (1983).

EXAMPLE. Consider a utility which must add 5 mgd capacity to a treatment plant at a cost of $7,000,000, but with leak detection will only need to add 3 mgd. The energy cost for pumping the water saved is $0.08/1000 gal. Assume the leaks would have been found in 2 years on the average without detection. Estimate the benefits due to water conservation.

$$B = \$7,000,000\{1 - [1 - (2/5)]^{0.7}\} + 2(694 \text{ gpm/mgd})(0.08)525(2)$$

$$= \$4,895,000 + \$285,000$$

$$= \$5,180,000.$$

Since this capacity expansion would last the utility 10 years, the leak repair program would pay for itself if it cost less than $500,000 annually.

The cost of water saved is not the only benefit of a leak detection program. Damages caused by the leaks can be significant, especially in high-value districts or landslide-prone areas. Savings in legal fees alone from a single catastrophic break can pay for a leak repair program. A leak detection program can also improve customer relations by informing water users when there is a leak on the customer's side of the meter. It also impresses on the customer the value of conserving water. There is benefit in the form of savings in facility capacity from having the customer reduce wasted water on the customer's property. These cannot, however, be valued the same as leaks on the utility's side of the meter, since the customer is paying for the water.

7.5.   REPAIRING BREAKS AND LEAKS

Once a leak or break is located, the utility has several options including doing nothing, repairing the leak or break, replacing the pipe, or taking some other remedial action such as cathodic protection in conjunction with repair. Doing nothing is attractive when the repair job would be difficult (e.g., difficult access), the water is not doing any damage, water is plentiful and inexpensive, and/or the pipe is scheduled to be abandoned or repaired later.

In general, there will be two levels of possible repair: (1) stop the leak or break, and (2) take action to prevent future problems. The first level will almost always be reached while the second depends on an economic analysis which will be described in later sections.

The first step in repair is to excavate to find the exact location of the break. The geophone or correlator-type leak detection devices are very helpful in this respect, since water does not always surface at the exact location of the break or leak. The next step is to shut down the system so that repairs can be made. This is done by shutting off valves to isolate the area of the break or leak. In some cases the repairs can be made while system pressure is still high. For larger breaks, the pressure may need to be reduced to virtually zero in order to make repairs. Except for very small leaks in porous soil, a portable pump will be needed to dewater the excavation. It is possible to dewater a trench by opening a hydrant at a lower elevation than the break and sucking water out of this trench. This practice is to be avoided because it introduces contaminated water into the system. Whenever possible, it is best to leave water in the pipe and keep pressure slightly greater than atmospheric.

For breaks less than two feet long, repair clamps which wrap around the pipe as shown in Fig. 7.3 can be used. Other types of clamps are available to hold pipes together at joints and to repair broken bells and flanges.

For larger pipe diameters clamps are not available, so for small holes, patches are generally welded onto the pipes. For breaks where the pipe is crushed, or cracks are extremely long, the entire section of pipe must be replaced.

Repair clamps and fittings should be swabbed with a one percent hypochlorite solution for disinfection. If the main being repaired has remained full of water, it is not necessary to disinfect the line. If, however, the pipe was drained, it should be flushed, disinfected with chlorine, and flushed again before being returned to service. Procedures are described in AWWA standard C601-81—Disinfecting Water Mains.

In addition to making the repairs, the crew can take several steps to assist the engineer in determining or implementing remedial actions. The crew should: (1) take a soil sample at the location of the break, (2) examine the exterior of the pipe for corrosion, (3) take photographs of the break or leak and the vicinity of the repair so that the exact location can be pinpointed

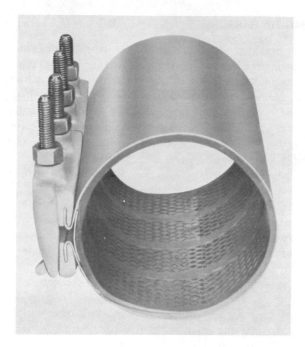

Fig. 7.3. Pipe clamp for leak repair. [Courtesy JCM Industries.]

later, (4) fill out a report describing the repair, and (5) install cathodic protection when it is deemed cost effective. While the first three steps are fairly obvious, steps four and five are described in some detail in the following paragraphs.

The engineer must know the cause of breaks and leaks if he is to make rational decisions on remedial actions to correct chronic problems. The engineer can determine the cause of problems if adequate descriptions of the breaks or leaks are provided. Fig. 7.4 gives an example form for recording the details of a repair. This form gives the engineer information needed to determine the type of break, its cause, and the cost of repair—all necessary pieces of information for selecting a strategy for future actions.

Corrosion of water mains can be reduced significantly by installation of cathodic protection. For very large pipes, an impressed electric current is usually the least expensive method of cathodic protection, but for most pipes sacrificial anodes which corrode instead of the pipe are less expensive. The major cost in installing sacrificial anodes is the cost of excavation and repaving.

If a trench is already open for pipe repair, it is a relatively simple matter to install a sacrificial anode in situations where corrosion is suspected as the cause of the break. Westerback (1982) reported that for this method to be

SANTA ROSA WATER UTILITY LEAK REPAIR REPORT

LOCATION:

MAP NO.:

DATE & TIME REPAIRED:

W.O. NO.

FOREMAN:

— DESCRIPTION OF DAMAGE —

WHAT PART WAS DAMAGED ?

☐ Pipe barrel ☐ Flange nuts, bolts, tie rods

☐ Joint ☐ Other -(describe): _____

☐ Valve _____

IN YOUR OPINION, WHAT CAUSED THE DAMAGE ? _____

TYPE OF BREAK:

☐ Split ☐ Crushed Pipe

☐ Hole ☐ Cracked bell

☐ Circumferential split

☐ Broken coupling

☐ Service pulled

☐ Cracked at corporation stop

☐ Gasket blown

☐ Other (describe) _____

☐ WATER MAIN ☐ SERVICE LATERAL SIZE: DEPTH TO TOP OF PIPE:

LOCATION OF LEAK: (circle number closest to leak)

PIPE MATERIAL:

☐ Galv. Iron ☐ Ductile iron ☐ Copper

☐ Black iron ☐ Cast Iron ☐ P.V.C.

☐ Steel ☐ A.C.P. ☐ Polybutylene

☐ Other _____

EXAMINE BROKEN EDGE OF CAST OR DUCTILE IRON PIPE:

Original thickness:

Min. thickness of good grey metal remaining:

Deterioration is on: ☐ Outside ☐ Inside

SIZE OF LEAK: (circle one) A B C

Is there evidence of previous leak repairs in some general area ? ☐ YES ☐ NO

No of previous leak repair clamps present

Last repair date (if known)

TYPE OF SOIL:

☐ Rocky ☐ Sandy ☐ Adobe

☐ Clay ☐ Hard pan ☐ Loam

EXISTING BEDDING:

☐ Gravel / Sand ☐ Native soil

☐ Pea gravel

IN YOUR OPINION, SHOULD PIPE BE REPLACED ?

☐ Yes ☐ No ☐ Do not know

IF YES, EXPLAIN EXTENT ON REVERSE SIDE.

— DESCRIPTION OF REPAIR —

DAMAGED PART WAS: ☐ Repaired ☐ Replaced

If replaced, what material was used ?

IF REPAIRED, WHAT REPAIRS WERE MADE ?

☐ Leak clamp ☐ Repacked valve ☐ Other (describe) _____

☐ Welded ☐ Recaulked joint _____

FILL IN THE FOLLOWING:

1. Street name, north arrow;
2. Draw in main and hydrants in shutdown area;
3. Show all valves closed and valve numbers;
4. Locate leak to nearest intersection or house with address. Show dimensions to property lines or street centerlines.

ATTACH THREE PHOTOS:

1. Straight down over leak or damage;
2. Close up of leak or damage;
3. Any other photo which you feel will help.

NOTE: If possible, tag and label any replaced part which you feel describes the general level of corrosion or deterioration and deliver part to water field office with this form.

WORK ORDER, LEAK REPORT, & PICTURES MUST BE SUBMITTED AS SOON AS POSSIBLE !

Fig. 7.4. Example of leak repair report. [Reprinted with permission of the City of Santa Rosa.]

successful, the anode should be installed whenever it is not completely clear that corrosion was not the cause of the break, since if there is some doubt as to whether corrosion is at fault, crews will prefer not to go through the trouble of installing the anode.

7.6. PROJECTION OF BREAK RATES

Knowing the future break rate for water mains is important for making decisions on whether to replace or repair pipes and for projecting costs of repair in future years. The break rate in future years is usually the current break rate with a slight increase over time, since as systems age the rate of breaks increases gradually. In this section the term *breaks* is used to refer to both pipe breaks and joint leaks. In projecting breaks, service line breaks and leaks should be considered separately because mains and service lines should be considered separately in economic analyses.

Discussing breaks in terms of total breaks in a system can be somewhat misleading, since the number of breaks will depend on the size of the system and time period under consideration. Instead, break rate (expressed as breaks per unit length of pipe per unit time) should be used. This enables the engineer to compare break rates between systems and time periods. For example, two pipes may have each experienced five breaks, but if one is 500 ft long and all the breaks occurred in one year, it is a much more likely candidate for remedial action than the other, 3000-ft-long pipe for which the breaks occurred over a 20-year period. In this text, units of breaks per year per mile are used and service leaks and hydrant leaks are not included.

Trends in break rate provide useful information on the causes of breaks and possible remedial actions. For example, if corrosion is the principal cause of breaks, then the rate will increase with time. On the other hand, if breaks are highest after severe winters, then frost penetration may be the factor that triggers breaks. Breaks due to impact or contact with other structures tend to coincide with construction activity in the area.

As stated earlier, the break rate in some future year is likely to be the current rate with some increase due to deterioration of the system with age. This statement is true only if the current year is a typical year. For example, if the current year is a very wet year in an area with expansive soil, the rate will be unusually high. Similarly, the rate in a year in which a listening survey is first conducted will not be typical. Break rates in a given system can fluctuate by a factor of ten from one year to the next, so an average of the previous several years is the most reliable indication of the current break rate upon which to base estimates of future break rates. In all of these analyses, the engineer must try to distinguish the date on which the pipe broke from the date on which the break was detected or the date on which it was repaired.

This is not always possible, since small breaks can exist for some time without being detected and utilities may wait to repair small breaks until a crew is free or the weather is more suitable.

Determining the rate of change of the break rate involves extrapolating trends in the break rate. This can be done graphically by plotting break rate versus time on semi-log graph paper, with the break rate on the logarithmic axis. This corresponds to fitting the data to the equation recommended by Shamir and Howard (1979):

$$J = J_0 \, e^{b(t-t_0)} \tag{7.6.1}$$

where

J = break rate in year t, breaks/year/mile
J_0 = break rate in year t_0, breaks/year/mile
e = 2.718
b = rate constant, 1/year
t = year
t_0 = base year.

This is not the only function which approximates historical break rate data, but it has generally proven to be a useful approximation. For reasonably small values of b (<0.10), this equation can be approximated by

$$J=J_0 \, (1+b)^{(t-t_0)}. \tag{7.6.2}$$

Eq. (7.6.2) sheds some light on the meaning of the rate constant b. It can be viewed as the rate of change of the break rate. For example, if $b = 0.03$, the break rate is increasing by 3% per year.

The engineer should not expect the break rate vs. time data to fall on a perfectly straight line when graphed. As mentioned earlier, such factors as weather and the rigor with which breaks are detected and recorded will vary from year to year. Thus, there will be considerable spread around the best-fit line. This spread can be reduced somewhat by grouping the data in sets of 3–5 years or using the moving average of the break rate plotted at the median year. This will tend to damp out noise in the data and enable the engineer to more easily distinguish trends due to time from those caused by other effects. If the effects of other phenomena, such as depth of frost penetration, can be quantified, they can be taken out of the data when it is plotted.

In some cases the data will be so sketchy that a few errors can result in significant errors in determining b. Walski and Pelliccia (1982) reported values of 0.021/yr and 0.014/yr for two types of cast iron pipes in Binghamton, N.Y., based on approximately 40 years of data. Clark, Stafford and Goodrich (1982) found a value of 0.09 for pipes which already had a previous break in Kenton County, Ky. In general, the value of b will be highest where corrosion is weakening pipes with time and lowest where other causes, which do not become worse with time, predominate. O'Day (1982), who did a considerable amount of work in Manhattan where contact with other structures was often the cause of breaks, reported break rates were not highly dependent of the age of the pipe.

Whenever the data are adequate, separate values of b should be calculated for different types of pipe and pipes layed under different conditions. Simply saying that pipe laid in 1920 is breaking at a higher rate than pipe laid in 1950 and calculating b based on the different rates is not adequate, because these pipes will be substantially different in material and laying conditions. The value of b should be interpreted as indicative of the change in break rate within a group of pipes, whether it is for the entire system or a specific type of pipe laid in a specific time period, rather than the rate of change between different groups of pipes.

7.7 COST OF BREAKS AND REMEDIAL MEASURES

In order to determine whether it is economical to replace pipes developing frequent breaks, or employ some other remedial measure as opposed to simply repairing the breaks, it is necessary to quantify the costs associated with both the breaks and the remedial measures. In this section, each of the major cost items involved with making sound economic evaluation of breaks is described. This information can be used as input to the economic evaluation techniques presented in the next section.

The cost of a break can be divided into costs for: (1) repair, (2) damages, (3) inconvenience to water users, (4) traffic delays, (5) health and safety effects, and (6) lost water. In rural or undeveloped areas, item (1) is the only significant item, and the others can be accounted for as a fraction of the first item. In developed, high-value areas, the cost to repair the break may be merely a minor item compared with the disruption of traffic and water service and damages. Each of the cost items is discussed separately in the following paragraphs.

The cost of repair depends on the type of break, the size of the pipe, the problems with shutting down the system, and the ease of access to the break. Specific cost items associated with repair are the crew, vehicle, equipment, clamps or sleeve, tools, repaving, supervision, and overhead. Most of these costs depend on the time required to make the repair. This, in turn, is highly dependent upon how long it takes to shut down the system, which in turn depends on the condition of valves that must be closed. If the valves have been properly maintained, they will be easy to locate and operate and will do an effective job of shutting off the flow. Unfortunately, for many systems, valves will be found broken if they can be found at all. This means the crew must search in ever widening loops to find operable valves, and service will be disrupted for more customers. In general, large transmission mains can be shut down more easily than mains in a tight grid, but larger valves are more likely to be inoperable than small valves. Repairs can be made most quickly if a sleeve or a joint clamp can be used. Welding a patch on the pipe or recaulking an old lead joint generally takes more time, while replacing entire sections of pipe is usually the most time consuming.

Once the time for repair is estimated, the cost can be determined by multiplying the time by the cost of a crew (remembering that breaks usually occur at night or on weekends, when overtime rates apply), compressor, backhoe or other excavation equipment, pumps, and vehicles per unit time. These costs will vary depending on the practices of the utility, but each utility should have data on the costs.

The costs of clamps, sleeves, and replacement pipe should be easily obtainable from the utility. The cost of hauling the pipe from storage will be a major item for large-diameter pipe.

Paving cost can usually be estimated on a dollars per square foot times area to be paved basis. If a temporary patch is installed when the break is repaired and a permanent patch is placed later, both costs should be included.

Two minor cost items which are difficult to quantify, but should nevertheless be included in an estimate, are overhead and supervision. The overhead cost should account for such items as space used for storage of repair equipment and a fraction of clerical, support, and utility costs attributed to break repair. Fringe benefits for the crew may be included in the labor cost or overhead, depending on the accounting practices of the utility. Time for supervision and inspection may need to be included if it is not generally considered in crew cost. In some cases, it may be best to include supervision and overhead as simply a fraction of the sum of the other costs.

As an example of repair costs, the U.S. Army Corps of Engineers, Buffalo District (1981), developed repair costs for the City of Buffalo as a function of pipe diameter for diameters ranging from 4 to 48 in. For example, for a 12-in. pipe the crew cost is $800, equipment is $275, sleeve is $80, paving is $355, and tools are $120, giving a total cost of $1630 in 1981 dollars. The relationship can be approximated by

$$\text{cost} = 600 \ D^{0.40} \tag{7.7.1}$$

where D = diameter, in.

This function represents typical costs based on practices for a single utility, and will vary for different utilities and for different breaks within a single utility.

When breaks occur in an area where there is little or no damage or interruption of service due to the break, then these costs can be accounted for by adding a percentage of the repair cost to the repair cost to obtain the total cost. Usually 20–50 percent is adequate. If the break occurs in an area in which these other costs are of the same size or larger that the repair cost, more effort needs to be expended quantifying them.

The costs of damage depend on the location, the size of the pipe, and the type of break. When a large pipe ruptures in a high-value district, the cost of the damage can run into many thousands of dollars. In an area with deep basements and buried utilities, damage estimates of several thousand dollars can be used. In areas with subways and significant underground activity, damage in the millions of dollars is possible, as occurred in New York City in August 1983. The cost of damages from future breaks should be based on estimates of damages from past breaks for that utility. For example, in New York City (U.S. Army Corps of Engineers, New York District, 1981), between 1963 and 1976 there were 205 claims filed each year for an average of $16,000 per claim. Of these, 125 were settled each year at an average of 29.1 percent of the claim. While costs in New York are generally higher than in most cities, claims are usually only filed for significant amounts of damages, so it is fairly safe to assume that at least as much damage went unclaimed, having been covered by insurance policies or simply written off as a loss by property owners.

While any major break is being repaired, service is interrupted in the vicinity of the break. If gate valves are functioning properly, this area can be confined to the block in which the break occurred. Otherwise, a fairly large area may be involved. A lower limit on the value of the water not delivered to the customer due to the interruption can be calculated as the volume of water not used times the unit price. This is referred to as a lower limit, since customers are willing to pay a high price for the first unit of water delivered to their household. This difference between willingness to pay and actual price is referred to as *consumer surplus,* and can be quite large. If the delay is unannounced, so that the customers cannot stock up on water, they will be willing to spend a considerable amount on bottled water during the interruption.

Even more serious impacts occur when a break forces an industry or commercial establishment to close. This cost should be related to the value of lost product.

Since most water mains are located under roads, repair of breaks often involves disruption of traffic and use of safety barriers and flag men.

Highway engineers have developed methods to evaluate the costs of such disruption. They can amount to as significant fraction of the total cost.

During a break occurrence, there is the risk that low pressures will cause water which has been contaminated around the break to be drawn back into the distribution system. There is also a loss in fire fighting capacity in the vicinity of the break. These problems occur rarely and are virtually impossible to quantify.

The value of lost water from a major break only becomes significant if the break is not repaired within a few hours. For example, a 100 gpm break of water valued at $1.00/1000 gal would cost the utility $6.00/hr. This represents a fairly large break, so this cost would only be significant if the break were large and went unrepaired for more than a day. For breaks in which the water finds an underground path to the nearby sewer or watercourse, the value of lost water could be the major cost item.

All of the above costs can be summed to give the cost of the break. This cost must be compared with the cost of some remedial action undertaken to prevent breaks. The most common include replacement of pipe, cathodic protection, and installation of restraints. The cost of each of these measures is discussed in more detail in the following paragraphs. A major cost item in any of these remedial measures is the cost of excavation and repaving. If repairs can be undertaken as part of routine repaving work, sewer improvements, or subway installation, the cost of the remedial measure can be reduced significantly. This is true for both direct installation cost and indirect costs related to noise, street closing, etc. Therefore, in calculating the cost of remedial measures, the cost will be different if the measure is undertaken by itself as opposed to being part of a general rehabilitation project.

Most utilities have considerable data on the cost of installing new pipe in their systems. If only a few sections of pipe are required, the unit costs will be significantly higher than that of a major addition to the system because the cost of mobilization to do the work will be relatively large. Whenever possible, it is best to schedule advance replacement of pipes (i.e., replacement to prevent future problems) as part of a large project rather than replace a few sections of pipe at a time. Differences in pipe cost between congested urban areas, where weak pipes can cause the most damage, and undeveloped areas must be considered (i.e., do not use the same unit cost for a given size pipe in both areas).

While most utilities have good data on the cost of new pipe, most do not have data on the costs of cathodic protection to prevent corrosion. Cathodic protection can be applied using either sacrificial anodes or an impressed dc current. With sacrificial anodes, another metal is buried near the pipe. This other metal, usually magnesium or zinc, will act as the anode in the corrosion cell and corrode instead of the pipe. To install sacrificial anodes, it is necessary to excavate to the depth of the pipe at various locations along the pipe route. The costs will therefore include anode, excavation, and repaving. Some allowance should also be made for periodic inspection and eventual replacement of the anodes after roughly 20 years, depending on the application.

In the case of an impressed electric current, a rectifier is needed to convert ac current to the required dc current, and to regulate voltage. Electrodes, usually made of graphite, are buried along the pipeline. Wires must be connected between the electrodes, the rectifier, and the pipe. Impressed current cathodic protection is normally most economical for pipes requiring significant current (generally large pipes) in highly corrosive soils. Impressed currents have the adverse side effects of accelerating corrosion of metal structures in the vicinity of the pipeline. In remote areas, the cost of bringing electrical power to the pipeline may make impressed current protection uneconomical.

If cathodic protection using sacrificial anodes is to be installed at the time a break is being repaired, the cost of cathodic protection can be decreased significantly, since only the cost of the incremental excavation and labor and the cost of the anode itself need be considered. Similarly, if cathodic protection is installed immediately before a street is to be repaved, the paving and inconvenience costs can be reduced dramatically.

Joint leaks, as opposed to breaks in the pipe wall, are often due to inadequate restraints at bends or fittings, and installation of a joint clamp can often remedy the cause as well as stop the leak. While the excavation equipment and crew are at the site, it may be economical to install joint clamps for several joints, upstream and downstream of the bend or fitting, if the thrust blocking is considered inadequate. Tie rods can also be installed as a remedial measure or the original thrust block can be supplemented if inadequate restraint proves to be the problem. If a joint leak is found near a bend or fitting and the leak is repaired, it is worthwhile to listen for other leaks in the immediate area while the crew is still at the site.

Estimating the costs of a break or some remedial measure to prevent breaks involves essentially the same techniques as construction estimating except that many of the costs are intangibles or are dependent on parameters that are not known at the time decisions must be reached. Nevertheless, the economic decisions discussed in the following section depend heavily on the cost estimates used, so these estimates should be as well quantified and documented as possible.

7.8. ECONOMIC EVALUATION OF BREAKS

The decision to take remedial action to prevent further pipe breaks should be based on expected savings as compared to the cost of allowing the pipe to continue breaking. Replacing pipes before they can do great damage is sometimes referred to as *advance replacement*. In this section economic criteria for selecting pipes for remedial action are derived. The primary criterion is that a pipe should receive remedial treatment if the present worth of the cost for the remedial action is less than the present worth of the costs of future breaks that will occur if the remedial action is not taken. The remedial action most often considered is replacement of sections of pipes which have been prone to breakage.

To perform the required economic analysis, data on the cost of breaks and remedial actions and some means for estimating future breaks, as presented in earlier sections, are required. The analysis should be made on a pipe-by-pipe basis rather than applying the results of a single evaluation to an entire group of pipes, because the analysis will identify the "bad actors" in the system.

The economic criterion stated above can be written in equation form as

$$pw_r < pw_b \qquad (7.8.1)$$

where

$$pw_r = \text{present worth of remedial measure}$$
$$pw_b = \text{present worth of future breaks.}$$

Generally, one is calculating the cost of installing replacement pipe in the base year or the first year thereafter in which the street is scheduled for repaving so there is only a single installation cost. The exceptions to this are cathodic protection by impressed current (where there is a continuing energy cost) and cathodic protection by sacrificial anodes (where there is a replacement cost for the anodes after 20–50 years). The present worth of cost of the remedial measure is, therefore,

$$pw_r = C_r + C_o/\text{crf} + C_2/(1 + i)^{tr} \qquad (7.8.2)$$

where

C_r = cost of installing remedial measure, $
C_o = operation and maintenance cost of
     remedial measure, $/yr
crf = capital recovery factor
C_2 = cost to replace remedial measure, $
i = interest rate
tr = time until remedial measure is to be
     replaced, yr.

In most problems, the operation and maintenance and replacement costs are negligible in comparison to the initial installation cost, so it is possible to use

$$pw_r = C_r. \tag{7.8.3}$$

A present worth factor is not applied to the cost of the remedial measure since it is considered to be installed in the base year. Defining the base year for the economic analysis as the year in which the remedial measure would be installed (if selected) simplifies the analysis. However, this means that the base year may be different for different pipes.

The present worth of future breaks is the sum of the present worth of expected breaks in future years. The probability of a break occurring in year j can be given by

$$J_j = J(1 + b)^j \tag{7.8.4}$$

where

J = break rate in base year, break/yr/mi
J_j = break rate in year j, break/yr/mi
b = rate constant for increase in breaks, 1/yr.

The present worth of breaks in year j can therefore be given by

$$pw_{bj} = \left(\frac{1 + b}{1 + i}\right)^j C_b J \tag{7.8.5}$$

where pw_{bj} = present worth of breaks occurring in year j, \$. The sum of breaks occurring in all years included in the planning horizon is

$$pw_b = JC_b \sum_{j=0}^{n} \left(\frac{1 + b}{1 + i}\right)^j \tag{7.8.6}$$

where n = number of years in planning horizon. If $b < i$, the exact value of n is not important as long as it is fairly large (>20 yr). This will almost always be the case.

Rather than substituting Eq. (7.8.6) into Eq. (7.8.1) in its current form, it is simpler to replace the summation in Eq. (7.8.6) with an integration and evaluate the integral

$$pw_b = C_b J \int_0^n \left(\frac{1 + b}{1 + i}\right)^x dx. \tag{7.8.7}$$

The integral can be evaluated to give

$$pw_b = C_b J \left[\left(\frac{1+b}{1+i}\right)^n - 1\right]\bigg/\ln\left(\frac{1+b}{1+i}\right) \qquad (7.8.8)$$

where $\ln$ = natural logarithm. In the above equation, the continuous equivalent to the discrete interest rate should be used, but as stated earlier for b, $(1+i)^x \approx e^{ix}$, whenever i is small (<0.15) so the difference between discrete and continuous interest are small. Substituting back into Eq. (7.8.1) for the present worth of breaks gives

$$C_r < C_b J \left[\left(\frac{1+b}{1+i}\right)^n - 1\right]\bigg/\ln\left(\frac{1+b}{1+i}\right). \qquad (7.8.9)$$

There is a break rate J^* that, when substituted for J in the above inequality, will make it an equality. This is called the *critical break rate*. If the current break rate exceeds this critical rate, then the pipe should be replaced or some other remedial measure be installed, because the cost of future breaks will exceed the cost of replacement or other remedial measures. This critical break rate can be given by the following equation:

$$J^* = \frac{C_r \ln\left[(1+b)/(1+i)\right]}{C_b\{[(1+b)/(1+i)]^n - 1\}} \qquad (7.8.10)$$

where J^* = critical break rate, break/yr/mi. In most cases $b < i$ and n is large, so that the term containing n is essentially zero. Thus, Eq. (7.8.10) reduces to

$$J^* = \left(\frac{C_r}{C_b}\right)\ln\left(\frac{1+i}{1+b}\right). \qquad (7.8.11)$$

When the break rate is not changing significantly with time, this equation can be further simplified to

$$J^* = (C_r/C_b)\ln(1+i). \qquad (7.8.12)$$

The above equations should be applied to each pipe in the system, one at a time. That means that C_r, C_b, J, J^*, and b should be determined for each pipe. Usually, however, it is better to calculate a typical J^* for a group of pipes. If the break rate is close to J^*, then the pipe is labeled a questionable pipe and precise determinations of C_r, C_b, and b can be made. These methods are discussed in greater detail in the following section. For large systems, it becomes useful to develop a data base system to keep track of pipe breaks.

For some short lengths of pipe, the break rate can appear to be very large even if there has only been one break. For example, if only three years of data are available for a 50-ft pipe with one break, the break rate is 35 breaks/yr/mi. This would seem to make the pipe a very likely candidate for replacement. In such a case, the engineer should examine the cause of the break to determine if it is part of a trend or an isolated incident.

While the criteria given in this section provide guidance on selection of pipes for remedial action, the rules given should be applied with a good deal of engineering judgment. There is considerable uncertainty in estimating the indirect costs of a break and the rate of increase of breaks. In marginal cases, the engineer should perform a sensitivity analysis to determine what the results would be if the damage costs were reduced by say 50% or the rate of increase of breaks was actually twice a high. Budget constraints must also enter the decision-making process, as sometimes there may simply not be enough money for advance replacement. A good economic analysis can be helpful in some instances in demonstrating that the utility cannot afford not to replace pipes.

7.9. BREAK INFORMATION RECORD KEEPING

To make sound decisions on water system maintenance, an engineer must have data on past breaks and pipe characteristics in a usable form. If information on breaks is recorded promptly and precisely and stored in a way that it can be retrieved easily, the engineer can concentrate on making decisions rather than on hunting for data.

The traditional way to keep track of breaks is to set up a file containing data on repair work, including such information as type of break, suspected cause, and cost of damage. A map with different-colored pins to indicate the type and location of breaks provides a great deal of the information needed to identify pipes which are good candidates for possible remedial measures. Such manual information storage and retrieval systems are adequate for small utilities, but can be unmanageable for larger systems.

In recent years, advances in computer technology have made it possible to develop automated information storage and retrieval systems. Lane and Buehring (1978), Curtiss and Lohmiller (1983) and O'Day (1982, 1983) have developed computerized systems for storing data on the condition of the distribution system for use in later decision making. O'Day's approach, which he refers to as *geoprocessing*, makes use of DIME (Dual Independent Map Encoding) geographic data base file from the U.S. Census Bureau. This data base contains data on individual street networks for over 200 cities in the

United States. The DIME file provides a base map on which information on the water system can be stored.

There are two types of information that must be stored: (1) a description of the system, whether it be a list of pipe segments, a map with pipe characteristics included, or a computerized data base; and (2) a description of breaks and repairs, whether it be a file folder in an engineer's desk or a computer data base. The pipe data should contain a description of each pipe, including the street name, beginning and ending identifiers (usually addresses), diameter, length, type of pipe, date laid, coordinates, pipe identification number (or node number of beginning and ending nodes) and type of joints. Other useful data may include a list of hydrants, valves, or services along the line, indexed so that this information can be found by cross reference with the files. Simply stating the type of pipe as "cast iron," for example is usually inadequate, since such pipe comes in several thickness classes and has been produced by several processes over the years. A better identification would be to identify a PVC pipe, for example, by its pressure rating and outer diameter, since such pipe comes in several sizes and thicknesses for a single nominal diameter. If the results of pipe roughness tests are available for the pipe, this information can be stored with the pipe description, as the decision to clean and line a pipe must also take into account whether the pipe is structually sound. Similarly, the decision to replace a pipe may depend on whether that pipe would need cleaning and lining if it is not replaced.

The break data file should contain, in addition to an identifier of the pipe which broke, data on the type of break, date of repair, date of detection, estimated date of break, type of break, estimated flow from break, cost of repair, estimate of damages, and an overall description of the condition of the pipe. The file should essentially contain all the information included on a leak or break report, with the exception of photographs (in the case of a computerized file). By linking the data base with computer graphics packages, it is possible to prepare maps of historic break locations.

Establishing a sophisticated information retrieval system will not be very helpful to a utility if the utility does not make an effort to maintain the system. This means continually updating the files with correct information. O'Day (1983) estimated that a water main data base system costs $50–100 per mile to establish and $10–20 per mile per year to maintain properly. While such a system will most likely pay for itself, the utility must remember that, if its initial investment is to pay off, it must make a long-term commitment to maintaining the system.

Any data base system is only as good as the information recorded in the field that will eventually end up in the system. A standard form for recording the location, cause and type of repair is essential for an analysis of breaks whether it be done manually or with a computer. Fig. 7.4 shows a good example of a leak repair report which provides necessary data for making decisions on remedial measures.

7.10. LOST AND UNACCOUNTED-FOR WATER

Not all of the water produced by a utility is sold to customers or even metered at the point of use. However, the terminology generally used to categorize water use often leads to some confusion, especially among the general public, as to the difference between lost and unaccounted-for water. Unaccounted-for water refers to the difference between the water produced (less water used within the treatment plant or plants) and water sold (or given) to users through metered connections. Therefore, the sum of water that is given away (but metered) plus the unaccounted-for water can be used to estimate total non-revenue producing water. Lost water, on the other hand, is water that leaves the system through leaks and other unauthorized uses. The difference between unaccounted-for water and lost water indicates that portion of the unaccounted-for water is actually put to productive use in fire fighting, main and sewer flushing, street cleaning, and under-registration of meters.

Unaccounted-for water and lost water are usually expressed as percentage of total production, but comparison of values from one system to the next is often misleading. For example, some utilities do not meter any public uses, others include an estimate of fire fighting and other public uses as accounted-for water, while still others meter only a portion of water users. The following paragraphs represent a suggested set of definitions which will hopefully make comparison of values between utilities more meaningful.

Before any attempt to quantify different types of use is made, water production at the treatment plant must be determined accurately, otherwise estimates of the other uses as fractions of water produced will be meaningless. The water produced should be defined as the water that actually leaves the treatment plants, since a considerable amount of water is used at the plant for purposes such as filter backwashing, and a significant volume of water may leave the plant as sludge or evaporate from sludge drying beds. Therefore, raw water taken into the plant should not be equated to production.

Table 7.1 provides a consistent set of definitions for alternative types of water use. At each level of the table every drop of water produced is considered in one and only one category. While the table may seem somewhat simple, it provides a consistent set of definitions with which to describe water use. The term *unaccounted-for* water is conspicuously absent because of the problems that exist in deciding which of the unmetered authorized uses contribute to that category.

From Table 7.1 it is clear that the key to evaluating the condition of a distribution system is the sum of the unauthorized use and leakage. However,

Table 7.1.   Types of Water Use.

I. Metered
   A. Sold
   B. Unsold
II. Unmetered
   A. Authorized
      1. Unmetered public buildings
      2. Unmetered customers
      3. Fire fighting and testing
      4. Flushing and street washing
      5. Net under-registration of meters
      6. Other authorized unmetered uses
   B. Lost water
      1. Unauthorized use
      2. Leakage

calculating this quantity can be quite complicated, because a significant portion of unmetered water may be put to authorized use. This difficulty can be overcome by: (1) metering public uses; (2) making rational estimates of fire fighting use based on duration of fires and tests; (3) eliminating unmetered customers, even if their water will be free or charged at a flat rate; and (4) regularly replacing and testing customer meters to reduce the amount of under-registration and, at the same time, provide data on how to make reasonable estimates of underregistration.

Estimates of each type of water use given above should be based on annual average use, since the percentages in each category will vary slightly from day to day. For example, in a community which does not meter watering of municipal parks, unmetered authorized use may increase significantly in hot, dry weather.

A utility with a reasonably tight distribution system can expect to keep lost water to less than 10 percent of production. This will be considerably easier to do for a utility that has large flow spread over a small geographic area than for a rural system which has a low flow with a large number of users. For example, a pipe in an urban utility may have a flow of 200,000 gal/day, while a utility in a sparse rural area may have only 10,000 gal/day. If 1000 gal/day/ mile is considered unavoidable leakage, the rural utility may appear to have excessive losses even though it actually has only the normal unavoidable loss. For this reason, it may be necessary to subtract unavoidable losses from lost water in order to make meaningful comparisons of the percentage of water lost in different systems (i.e., divide II.b.2 into avoidable and unavoidable leakage).

Overall, the key to keeping track of water use in a system is to meter all uses and to keep the meters in good operating condition.

7.11. APPURTENANCES

Water distribution appurtenances include hydrants, valves, and fittings. In most utilities these appurtenances are not used often and require little attention, but unfortunately they sometimes receive no attention at all. This can lead to disastrous results when a hydrant is found to be inoperative during a fire or a valve cannot be shut off to prevent flooding resulting from a break.

The benefits of a maintenance program for valves and hydrants are difficult to quantify because often a valve can be broken for years without causing any trouble. Broken valves prevent effective shutdowns after a break and make routine repair work difficult. Broken hydrants hamper fire fighting, and hydrants in poor condition can reduce the credits a community receives in fire insurance ratings. Similarly, the system can receive credits based on the frequency of inspection.

The costs of a maintenance program can be calculated fairly easily. There should be a two-person crew with vehicle, tools, paint, and lubricant to perform tests of valves and hydrants and paint hydrants. When major repair work is found to be required, this crew would request the assistance of a full-size maintenance crew, equipped with spare parts and necessary tools for major repairs. The details of hydrant inspection and testing are given in AWWA Manual M17, while valve inspection and testing is described in AWWA Standard C500-81.

REVIEW QUESTIONS

1. How does corrosion cause pipe breaks?
2. Why does waterhammer lead to leaks at joints near bends?
3. Why is the minimum night ratio a good indicator of leakage?
4. Why is listening for breaks less effective for plastic pipes than metal pipes?
5. How can an individual tell if a service line leak is on the customer's or utility's side of the meter?
6. If done properly the first time, why is it less expensive to conduct a water audit the second time?
7. Why is the price of water not a good indicator of the benefits of reducing leakage?
8. Distinguish between long-run and short-run cost savings due to leak reduction.
9. List the information that should be included in a report of a break.
10. How does cathodic protection work?
11. If $b = 0$ in Eq. (7.6.1), what is the implication on the break rate?
12. Why is age not necessarily a good indication of whether a pipe should be replaced?
13. In what kind of areas would the repair cost be the major cost of a break? In what area would the damage cost be largest?
14. Why is it not completely correct to estimate the cost of inconvenience to a

consumer due to interruption in service by simply estimating the value of water not consumed?

15. Why does impressed current cathodic protection become less costly than sacrificial anodes for large current demands?

16. Why is the base year for economic decisions regarding remedial measures for prevention of breaks taken as the year in which the measure would be implemented?

17. What is the meaning of J^* in Eq. (7.8.10)?

18. List some of the information that should be stored in a file describing: (1) a section of pipe, (2) a pipe break.

19. Why is there confusion about using the concept of unaccounted-for water in assessing the adequacy of a distribution system?

20. Name several ways of reducing unmetered authorized use.

21. Name the AWWA Manual that describes fire hydrant testing and inspection.

PROBLEMS

1. How many breaks must occur in a 2000-ft water main in a 10-year period to make it economical to replace that pipe rather than continue to repair breaks? The cost to replace the pipe is $60/ft and the cost to repair a break is $1800. Other break costs average $3000. Use an interest rate of 10% and $b = 0.02$. What if $b = 0$? (*Ans.* 25 and 32 breaks.)

2. A utility with 125 miles of water mains which produces an average flow of 12.12 mgd has 8.71 mgd of metered sales, 0.37 mgd of metered use at public buildings, and 141 hours of fire fighting at an estimated 2000 gpm. Older meters are under-registering by an average of 8%, but an average meter is estimated to under-register by 2%. What is the percent lost water? Suppose unmetered public use is estimated at 1.6 mgd. Recalculate the lost water. Now suppose that 1000 gal/day/mi is attributed to unavoidable lost water, what is the percent avoidable lost water? (*Ans.* 23, 10, 9%)

3. As part of a water audit, a district within a utility has a minimum nighttime flow of 1200 gpm and an average daily flow of 2000 gpm. Find the minimum nighttime ratio. Now, suppose there is an industry in the district with an average use of 800 gpm and night use of 600 gpm and there is a 40-ft storage tank in the district in which the water level is rising at the rate of 2 ft/hr at the time of the minimum nightly flow. Recalculate the minimum nighttime ratio using this additional information. (*Ans.* 60, 24%.)

4. Suppose a utility buys leak detection equipment for $3000 and uses a two-man crew for $500/day for 180 days per year to detect leaks. During the year they found 70 minor leaks with an average flow of 4 gpm. These leaks cost an avearge of $100 to repair. They also find 20 major leaks having an average flow of 20 gpm and repair cost of $2500 each. Given a long run unit cost of water of $0.50/1000 gal and assuming a leak would have gone on one year longer without the detection program, is the program worthwhile? What if the leaks would have gone on for 6 months? 2 years? (*Ans.* yes, no, yes.)

REFERENCES

American Water Works Association, 1980, AWWA Standard for Gate Valves, 3 through 48 in. NPS for Water and Sewage Systems, C500-80.

AWWA, 1981, AWWA Standard for Disinfecting Water Mains, C601-81.

AWWA, 1980, Installation, Field Testing, and Maintenance of Fire Hydrants, AWWA M17.

AWWA Task Group 2850-D, 1969, "Replacement of Water Mains," *J. AWWA*, Vol. 61, No. 9, p. 447.

Boyle Engineering, 1982, Municipal Leak Detection Program Loss Reduction—Research and Analysis, State of California Department of Water Resources.

Clark, R.M., C.L. Stafford and J.A. Goodrich, 1982, "Water Distribution Systems: A Spatial and Cost Evaluation," *J. ASCE WRPMD*, Vol. 108, No. WR3, p. 243.

Cole, G.B., 1980, "Leak Detection—Two Methods That Work," *AWWA Seminar on Operational Techniques for Distribution Systems*, Los Angeles, Calif.

Curtiss, J.F. and P.A. Lohmiller, 1983, "Computerized Distribution Record—CADD Paves the Way," *AWWA Distribution Symposium,* Birmingham, AL, p. 65.

Heim, P.M., 1979, "Conducting a Leak Detection Search," *J. AWWA*, Vol. 71, No. 2, p. 66.

Kingston, W.L., 1979, "A Do-It-Yourself Leak Survey Benefit-Cost Study," *J. AWWA*, Vol. 71, No. 2, p. 70.

Lane, P.H. and N.L. Buehring, 1978, "Establishing Priorities for Replacement of Distribution Facilities," *J. AWWA*, Vol. 70, No. 7, p. 355.

Laverty, G.L., 1979, "Leak Detection: Modern Methods, Costs and Benefits," *J. AWWA*, Vol. 71, No. 2, p. 61.

Moyer, E.E., J.W. Male, I.C. Moore, and J.G. Hock, 1983, "The Economics of Leak Detection," *J. AWWA*, Vol. 75, No. 1, p. 28.

O'Day, D.K., 1982, "Organizing and Analyzing Leak and Break Data for Making Main Replacement Decisions," *J. AWWA*, Vol. 74, No. 11, p. 588.

O'Day, D.K., 1983, "Geoprocessing—A Water Distribution Management Tool," *Public Works*, Vol. 114, No. 1, p. 41.

Shamir, U., and C.D.D. Howard, 1979, "An Analytic Approach to Scheduling Pipe Replacement," *J. AWWA*, Vol. 71, No. 5, p. 248.

Siedler, M., 1983, "Winning the War Against Unaccounted-for Water," *ASCE Conference Urban Water*, 84, Baltimore, MD.

U.S. Army Corps of Engineers, Buffalo District, 1981, *Urban Water Study, Buffalo, NY.*

U.S. Army Corps of Engineers, New York District, 1980, *New York City Infrastructure Study*, Vol. 1, *Manhattan.*

Walski, T.M., 1983, "The Nature of Long Run Cost Savings Due to Water Conservation," *Water Resources Bulletin*, Vol. 19, No. 3, p. 489.

Walski, T.M., and A. Pelliccia, 1982, "Economic Analysis of Water Main Breaks," *J. AWWA*, Vol. 74, No. 3, p. 140.

Westerback, A.E., 1982, "Cathodic Protection of Water Mains," *Public Works*, Vol. 113, No. 4, p. 49.

8 | TESTING WATER DISTRIBUTION SYSTEMS

8.1. INTRODUCTION

Some fairly powerful analytical tools for making decisions on water distribution systems have been presented in the preceding chapters. Because use of these tools involves a good deal of effort, an engineer can occasionally lose sight of the need to gather good quality data as input to, for example, a modeling effort or an analysis of whether to clean and line a water main. However, the temptation to use "typical literature" values for such quantities as pipe roughness must be avoided because, for virtually any real system, there is significant deviation from these "typical" values. Fortunately, it is fairly inexpensive and relatively easy to measure the most important quantities accurately.

Failing to make field measurements of an existing system when designing expansions or contemplating rehabilitation often leads either to a system with inadequate capacity or needless expenditure brought on by overdesign. The cost of gages and the time spent in conducting the necessary tests represents a small investment in good management that can pay large dividends. "Saving money" by not conducting tests as part of a modeling study or the design of improvements is like saving money on automobile maintenance by not putting oil in the engine. In fact, tests should be conducted on a regular basis, since conditions can change within a system from one year to the next.

In this chapter, the quantities which can be directly measured in the field, such as pressure, flow in pipes, and hydrant discharge, are discussed first. This is followed by a description of how to calculate quantities such as pipe roughness or hydrant discharge at a given residual pressure. There are several options available for calculating virtually any quantity, and the correct decision will depend on site conditions, accuracy required, and available equipment.

8.2. PRESSURE MEASURING DEVICES

Since water needs to be delivered with adequate pressure, pressure measuring devices are essential parts of any water distribution system. Yet many utilities have only a rough idea of the pressure in their system and others rely solely on customer complaints to detect low pressures. Accurate pressure data are essential for calibrating water distribution models and for extrapolating the results of hydrant flow tests.

There are five types of pressure measuring devices: manometers, Bourdon gages, bellows gages, diaphragm gages, and electric transducers. (Some authors use the word *transducer* to refer to any device which converts energy from one system to another. In the water industry, *transducer* is generally reserved for devices which produce an electric output, and this convention will be used in this chapter.)

Because of the magnitude of the pressures found in water distribution systems, manometers are not practical for measuring line pressures, although they are useful in differential pressure measurements which are discussed later. Among the gages, the Bourdon tube mechanism is the most commonly used in water system applications. Bourdon tubes are relatively accurate, rugged, and inexpensive. Electric transducers can be even more accurate but are more delicate and require a power source. Thus, they are more appropriate for fixed installations than for field use.

The accuracy of Bourdon tube gages is generally expressed as a percentage of the full scale reading. Gages with 0.25% accuracy are called *test gages* and are fairly expensive and delicate. Gages with 0.5, 1, and 2% accuracy are called *grade 2A, A,* and *B gages,* respectively in accordance with ANSI Standard B40.1-1974. For simply checking the pressure at a point in a system, any inexpensive pressure gage will be acceptable, but if the pressure readings are to be used for model calibration, a Grade A gage or better should be used. Since accuracy is given as a percentage of full scale, it is important to select a gage that has a scale only slightly larger than the expected pressures to be encountered. For most utilities, a gage with a 100 psi scale is ideal, although some may need a gage with a higher range. Dials should have a minimum diameter of $3\frac{1}{2}$ in. and the dial should be graded in 1 psi increments. This helps to eliminate reading errors. Ideally, gages should capable of being read over a 270 degree arc. The connection most commonly used at the bottom or back of the gage is $\frac{1}{4}$-in. N.P.T. (National Pipe Thread) male.

Transducers having accuracies of 0.5% of scale and better are available. In purchasing such a gage, it is important that the accompanying electronic equipment (demodulators, recorders, indicators, amplifiers, cable, and transmitters) be matched precisely with the transducer and power source.

Recording devices are available for both gages and transducers. The gages are best for remote locations without power because they can be spring or battery driven. Recording devices for transducers are more complicated, but the transducer output can also be digitized and stored on tape or disk for later use. Since recording gages are generally used for taking measurements over long periods of time, they are usually not manned continuously. Thus, they should be used only in a secure area or placed in a vandal-proof case and locked onto the hydrant or pipe.

Permanently mounted gages are generally tapped directly into the pipe. For field testing the gages are usually mounted on the $2\frac{1}{2}$-in. hydrant outlets. An old hydrant cap can be tapped and a gage can be inserted in it for field use. A number of manufacturers make a pressure gage that is mounted on a lighter cap which is easier to work with than a heavy hydrant cap. One such gage is shown in Fig. 8.1. Some of the gages have a bleed valve to let air out of the hydrant. This type of valve is critical if the gage is to be used on wet barrel hydrants, since if such a valve is not opened when the hydrant is closed the gage will be driven off scale, damaging the gage.

In addition to having a gage mounted on a $2\frac{1}{2}$ in. hydrant cap, a utility should have on hand fittings which will make it possible to attach the gage to virtually any kind of threaded opening. The gage connected to a $\frac{3}{4}$-in. hose bib (Fig. 8.2) is one example of one fitting which is very helpful in checking

Fig. 8.1. Pressure gage on hydrant cap.

Fig. 8.2. Pressure gage on hose fitting.

pressures. If pressure is to be read at a point removed from the main, steps should be taken to insure that there be no flow between the main and the point at which the pressure is measured. Otherwise, the flow will result in a low reading. Such a reading would be useless for model calibration.

When measuring pressure at a hydrant, the hydrant valve should be completely open to insure that the drains in the barrel are closed. In most hydrants, the drains will remain open when the hydrant valve is partially open. This can undermine the soil around the hydrant.

8.3. DIFFERENTIAL PRESSURE

The gages discussed in the preceding section are used to measure gage pressure (i.e., pressure in excess of atmospheric pressure). In measuring head loss in a pipe or the pressure difference associated with a venturi or orifice meter, or the difference between the static and dynamic pressure legs of a pitot tube, the engineer must be able to measure small differences in pressure between two fairly large pressures. Measuring pressures using two gages and subtracting the readings will generally not yield accurate results (e.g., two readings may differ by only one percent but each may be accurate to only one percent).

In such a situation a device which measures differences in pressure can produce much more accurate readings than using two gages. There are three

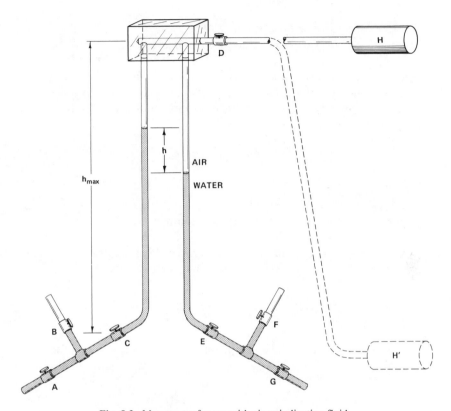

Fig. 8.3. Manometer for use with air as indicating fluid.

types of devices used to measure differential pressures: manometers, bellows-type pressure gages, and differential pressure transducers. Each of these is appropriate for some specific application and each has some advantages over the others. Features of the devices and guidance on selecting devices for specific applications are discussed in the following paragraphs.

A manometer operates by offsetting the differences in pressure between the two ends of the manometer by the different heights of manometer fluid on each side of the device. Given the manometer shown in Fig. 8.3, the difference in pressure can be given by

$$\Delta p = h(\omega_w - \omega_m) \qquad (8.3.1)$$

where

Δp = differential pressure, F/L^2

h = difference between height of columns of manometer fluid, L

ω_w = specific weight of water, F/L^3
ω_m = specific weight of manometer indicator fluid, F/L^3

In the case in which the indicating fluid is heavier than water, the specific weights of water and the indicating fluid should be switched. In this section, the formulas are given for a lighter-than-water indicating fluid. Dividing through by the specific weight of water gives

$$\Delta h = h(1 - SG) \tag{8.3.2}$$

where

Δh = differential pressure, L
SG = specific gravity of manometer fluid.

For air as a manometer fluid, the specific gravity depends almost completely on the gage pressure in the manometer, which can be given by the ideal gas law as

$$SG = 0.0012(1 + p) \tag{8.3.3}$$

where p = gage pressure in atmospheres. Substituting this back into Eq. (8.3.3) gives

$$\Delta h = h(1 - 0.0012p). \tag{8.3.4}$$

Note that 1 atmosphere is equivalent to 14.7 psi or 101 kPa.

The manometer in Fig. 8.3 uses a fluid that is lighter than water as an indicating fluid, as the U-shaped portion of the manometer points upward. If the manometer fluid were heavier than water, the U would point downward. The range of differential pressures measured by the manometer can be increased by making it taller or using an indicator fluid that has a specific weight considerably different from that of water. The manometer can be made fairly sensitive by using a fluid with a specific weight almost equal to that of water. Manometers can be made more sensitive when inclined. In that case the value of h in Eq. (8.3.1) should be divided by cos α, where α is the angle the manometer makes with the vertical.

Manometer indicating fluids should be immiscible in water and, for potable water systems, nontoxic. Knowing the height of the manometer and the maximum differential pressure reading that will be encountered, it is possible to select the manometer fluid by solving Eq. (8.3.1) for the specific weight of the fluid. Unfortunately, nontoxic fluids are only available in certain specific weights.

When air or some other gas is used as the manometer fluid, the density of the gas is often so small that it can be treated as zero for most practical problems with less than one percent error. Such an air-filled manometer is shown in Fig. 8.3. The container H on the manometer in Fig. 8.3 is used to store air when the manometer is placed under pressure. By raising or lowering the container, air or water can be introduced into the manometer so that the vertical columns can be roughly half full of water to give the maximum range. Valve D is closed when readings are taken.

Bellows pressure gages are somewhat more expensive and delicate than the Bourdon tube gages described in the previous section, and for field application usually require a case or mount. They can be purchased (with case, bleed valves, and manifold) from several manufacturers. Manifolds are useful because they enable the gage to be turned off without the problems that would arise if each side were turned off separately. A typical differential pressure gage with manifold and case is shown in Fig. 8.4.

Differential pressure transducers are available for the ranges commonly encountered in water distribution system work, and can be equipped with a wide array of indicators, transmitters, filters, and recorders. However, at present, there are no commercially available complete units designed specifically for field use for water systems, so the engineer must assemble (or have the manufacturer assemble) the unit from individual components.

Fig. 8.4. Differential pressure gage.

For use with a pitot tube, which will be described later, a differential pressure device should have a range of roughly 1 psi (27.7 in. water). This can be achieved with manometers, gages, and transducers. Manometers have the distinct advantage of measuring pressure differences in either direction. Gages and transducers can be purchased with zero at some point other than the end of the scale, but this means the accuracy is slightly reduced, since accuracy is usually related to full scale deflection.

In addition to the range, two pressure values are important in selecting a differential measuring device: line pressure and maximum overpressure. Line pressure is important in selecting the housing, hoses, and valves for the device. For water systems, a pressure rating of 200 psi will usually be adequate. Overpressures (i.e., differential pressures in excess of the range of the device) can damage the gage if adequate protection is not provided. Most gages have over-range protection devices which protect the gage at pressures up to rated line pressure, while transducers are usually only capable of withstanding overpressures of about twice the range. These overpressures may occur when a hose or fitting breaks, or when a worker opens the wrong valve or disconnects one port of the device while leaving the other under pressure.

Manometers are not damaged by overpressures, but overpressure can force manometer fluid from the manometer into distribution system. Fluid can also be drawn into the system if the manometer should ever fall over. Because of the potential problems caused by the toxicity of many manometer fluids, the engineer should be careful to select fluids which would be nontoxic if drawn into the distribution system. Air is an especially attractive manometer fluid because it is harmless and available everywhere.

All three types of device perform well in the laboratory, but only gages are available in complete kits, which consist of gage, carrying case, manifold, and hoses. In specifying hoses, it is important to be consistent with all connections (e.g., used $\frac{1}{4}$-in. N.P.T. connections with the female end facing the direction of the gage for all fittings). While transducers are fairly durable, the electronic equipment required for display and recording must be kept dry and spared any rough handling. The primary shortcoming of transducers for field use is the need for a power source. For an important set of measurements this equipment can be mounted in a van with a mobile power source, but this is quite costly. Some transducers are available for use with batteries. Glass manometers are fairly delicate, but plastic manometers can be purchased or fabricated quite easily. In making a manometer, be certain to use tubing of sufficient diameter ($>\frac{1}{4}$-in.) so that bubbles will not be trapped in the tubing.

In summary, transducers are best for permanent, indoor installations where a power source is available, as they can produce digital output which can be read, transmitted, and stored. Manometers are best for field applications, since they are inexpensive, durable, and accurate. For water distribution system applications, however, they should be used only with nontoxic indicator fluid. This means that the manometer may have to be extremely large for larger differential pressures. Gages can be purchased with wider ranges than manometers and are commercially available with carrying cases and manifolds. The indicators can be connected to spring driven chart recorders so that they can be used in some applications where neither transducers nor manometers would be suitable.

8.4. PRESSURE SNUBBERS

Water pressure in a distribution system does not remain constant, but rather fluctuates continually due to surges which arise whenever flow in the system changes even slightly. In most instances, these fluctuations are on the order of a few percent of pressure and cause little difficulty in obtaining accurate readings. However, when differential pressures are to be read to hundredths of a psi, these fluctuations in pressure can cause dials on gages and readings on indicators attached to transducers to fluctuate wildly.

Fluctuations can be dampened out using any type of fitting that restricts water from moving rapidly to and from the measuring device. Partially closing a valve between the measuring device and the main can dampen the fluctuations; however, there are special devices known as *pressure snubbers* which can be placed immediately ahead of the gage to dampen fluctuations. Snubbers slow down the movement of the device so that by the time the indicator begins to respond to a sudden positive surge, the surge will be over. Snubbers not only make it easier to read pressure devices, but they protect gages from damage due to excessive pressures.

Most pressure snubbers are made of a filter (usually stainless steel) with $\frac{1}{4}$-in. N.P.T. thread at each end. An alternative design consists of a piston which fits into a tube in such a way that a surge closes off flow to the pressure measuring device. It is possible to "oversnub" the pressure by using a too restrictive snubber, so that it may take several minutes for a device to respond to a change in pressure. Thus, snubbers must be carefully matched to the application. Oversnubbing can be especially annoying when one is measuring a change in pressure during a hydrant test which will only last a minute. Ideally, with a snubber in place, it should take approximately 10–30 seconds for the device to indicate the new pressure, given a sudden change in pressure.

A snubber is required on each connection of a differential pressure gage, and these snubbers should be as close to identical as possible. If not, it is possible for the indicator to appear to have reached steady state when actually one side is still moving (very slowly) to the correct reading. Fairly coarse snubbers should be used with a hydrant pitot gage (as discussed later) so that the test will not have to be run for an excessive length of time.

8.5. FLOW MEASUREMENT

For any study of a water distribution system, it is necessary to know flow in pipes and discharge from hydrants and hoses. There is a large group of devices which can be used for flow measurement, and the principles involved vary considerably. Some of the devices must be permanently mounted in the pipes (e.g., orifice plate, venturi meter) while others can simply be clamped on the outside of a pipe (e.g., ultrasonic and doppler meters).

In the following sections, devices for measuring flow in mains are discussed and methods for measuring discharge are presented. Except for master meters at the water source and interconnections with other utilities most systems do not meter flow in mains. Such information, however, is very useful for both water audits and model calibration. When selecting a flow meter, the engineer must consider: cost, ease of access to meter, continuous or occasional use, acceptable head loss, and need for recording devices.

The following description of flow meters was extracted primarily from Fath (1978), whose work serves as the basis for Table 8.1. For a more detailed reference see Miller (1983).

8.6. VENTURI METERS

Venturi meters are commonly used as master meters for water systems. They produce little head loss and are accurate over a wide range of flows. A venturi meter as shown in Fig. 8.5 operates on the principle that as the diameter of a pipe decreases the velocity head increases and, if the total head is to remain virtually constant, the static pressure must decrease. This change in static pressure between the sections with different diameters can be detected by a differential pressure device and is proportional to the flow squared, as given below:

$$Q = C_v A_2 \sqrt{\frac{2g\Delta h}{1 - (D_2/D_1)^4}} \tag{8.6.1}$$

where

Q = flow rate, cfs
C_v = velocity coefficient
A_2 = area of throat, ft^2
Δh = change in static head between cross sections, ft

g = acceleration due to gravity, ft/sec^2
D_1 = diameter of pipe, ft
D_2 = diameter of throat, ft.

In choosing a venturi meter, the engineer must select the ratio of diameters and static head change such that the differential pressure indicator will be correctly matched with the venturi over the entire range of velocities encountered. Differential pressure indicators can be purchased with square root scales so that the flow can be read directly from the indicator without the need to use Eq. (8.6.1). The coefficient C_v in the equation decreases for Reynolds numbers less than 10^5, so this direct readout cannot be used for small pipes with low velocity.

Venturi meters are fairly expensive and require straight upstream piping for roughly 10 pipe diameters. They are used only for permanent installations.

8.7.   ORIFICES AND NOZZLES

Either an orifice plate or a nozzle (Figs. 8.6 and 8.7) can be placed inside a pipe to produce a pressure difference which is proportional to the velocity squared and can be detected by a differential pressure measuring device. Both the orifice plate and the flow nozzle are considerably less expensive than the venturi meter and produce a greater difference in head than the venturi, so that a less sensitive differential pressure device can be used. Unfortunately, the orifice plate and nozzle produce greater head loss than the venturi meter and can trap sediment, especially in the case of the orifice plate.

Standard designs for orifice plates and nozzles exist so there is usually no need to calibrate such devices in place. Flow can be given by

$$Q = A_2 C \sqrt{2g\Delta h} \qquad (8.7.1)$$

where

A_2 = smallest cross-sectional area in the orifice or nozzle, ft^2
C = coefficient.

The coefficient C is a function of the ratio of diameters between the pipe and the orifice or nozzle and the location of the pressure taps, and is independent of Reynolds number for Reynolds numbers (based on full pipe velocity and diameter) greater than 10^5. If the taps are to be located at points in the pipe different from those on which C is based, the orifice must be calibrated in place. C values are higher for nozzles than for orifice plates with the same opening, so nozzles tend to be used for applications involving higher velocities.

Table 8.1. Summary of Water Flow Measuring Devices.

Device	Typical Max.	Flow Rangeability Max to Min	System Accuracy	Scale Characteristic	Relative Installed Cost	Typical Pressure Loss	Straight Upstream Pipe	Remarks
Orifice plate	Unlimited	4:1	±1% Max.	Square Root	Low–Med	50–90% of ΔP	Per ASME Standards 10–30 Dia	Most Popular expanded scale at high flow
Venturi	Unlimited	4:1	±1% Max.	Square Root	High	10–25% of ΔP	10–30 Dia	Low Pressure Loss
Flow nozzle	Unlimited	4:1	+1% Max.	Square Root	Med	30–70% of ΔP	10–30 Dia	Often used on High Velocity Applications
Flow tubes	75,000	4:1	±1% Max.	Square Root	Med	2–4% of ΔP	10–30 Dia	Very Low Pressure Loss
Rotameter; kinetic manometer	120,000	10:1	±3% Max.	Linear	Low	50–90% of ΔP	Per ASME Standards 10–30 Dia	Economical High Flow meas. Used with Orifice
Elbow meter	Unlimited	4:1	±5% Max.	Square Root	Low	Negligible (same as elbow)	25 Dia	Low Cost High Flow Meas.
Pitot (typical)	Unlimited	4:1	*±5–10% Max.	Square Root	Low	Negligible	20–30 Dia	Using multiple traverses and accurate pipe meas., accuracy may be ±1%

Type	Max.	Rangeability	Accuracy	Output	Cost	Pressure Loss	Straight Run Required	Remarks
Pitot (rotameter)	2,500	10:1	±3½% Max.	Linear	Low	Negligible	20-30 Dia	Direct reading Rotameter
Pitot (annubar)	250,000	4:1	±2% Max.	Square Root	Low	Negligible	7-25 Dia	Can be inserted during Flow
Propeller flowmeter	60,000	10:1	±2-5% Rate	Linear	Med-High	1-25″ Water	5 Dia	Usually used to totalize flow
Vortex shedding flowmeter	90,000	10:1	±1-2% Rate	Linear	Med	Up to 300″ Water	10-40 Dia	Reynolds No. >10,000 req'd. digital output.
Ultrasonic clamp-on flowmeter	300,000	10:1	±3-5% Rate	Linear	Low	Negligible	10-30 Dia Critical	Sensitive to pipe material and internal surface conditions.
Ultrasonic thru-wall flowmeter	Unlimited	25:1	±1% Rate	Linear	Med-High	Negligible	10-20 Dia	Flow Profile Sensitive
Electromagnetic flowmeter	Unlimited	25:1	±½% Rate	Linear	Med-High	Negligible	Negligible	No zero setting req'd. in ultrastable types. Direct reading range dial.

After a table by Fath, with permission of the author.

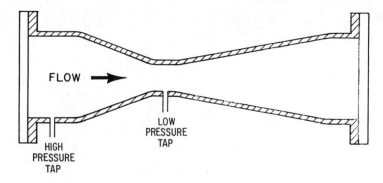

Fig. 8.5. Venturi flowmeter. [Reprinted with permission of J.P. Fath.]

EXAMPLE. Consider a 24-in. pipe which will carry a peak flow of 5000 gpm. Determine the range of a differential pressure gage (in psi) which will be required if orifices with openings of 18, 12, and 6 in. are used ($C = 0.5, 0.38$, and 0.36, respectively). What will the gage attached to the 12-in. orifice read when the flow is 1000 gpm?

Solve Eq. (8.7.1) for Δp (in psi) to give

$$\Delta p = \frac{0.43}{2g}\left(\frac{Q}{CA_2^2}\right)^2.$$

Substituting $Q = 5000$ gpm $= 11.1$ cfs and $g = 32.2$ ft/sec^2 gives

$$\Delta p = 1.33/(C^2 D_2^4).$$

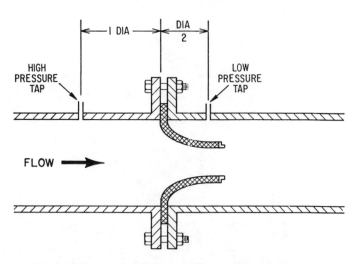

Fig. 8.6. Flow nozzle. [Reprinted with permission of J.P. Fath.]

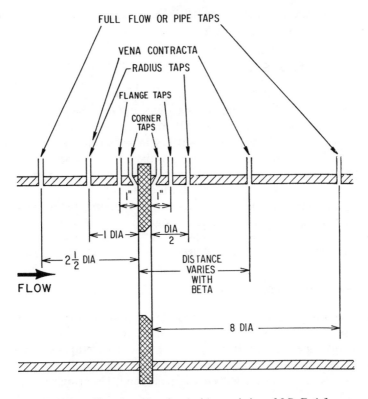

Fig. 8.7. Orifice plate. [Reprinted with permission of J.P. Fath.]

Substituting for C and D_2 and calculating Δp gives:

D (in.)	6	12	18
Δp (psi)	59.1	3.46	0.52

If the 12-in. orifice is used, the reading when the flow is 1000 gpm (2.23 cfs) is

$$\Delta p = 3.46 \, (1000/5000)^2 = 0.138 \text{ psi.}$$

If a pressure gage with a full scale of 5 psi was used, a reading of 0.138 psi would represent only 4 percent of full scale deflection. This is a problem with any of the gages for which flow varies as a square root of change in pressure. If the gage is to be able to record high flows, it will not be very accurate for low flows. It is therefore important to accurately anticipate the flows to be measured. When low flows predominate, but the meter must be able to measure rare high flows (and accuracy is important), compound meters should be used even though they are fairly costly. Fortunately, master meters at treatment plants and pumping stations operate at fairly constant flow rates.

8.8.   FLOW TUBES

Flow tubes (also called Dall or Dahl tubes) combine the low head loss of the venturi meter with higher pressure readings because the high-pressure tap faces upstream and, hence, senses total pressure instead of static pressure, as shown in Fig. 8.8. There are several devices on the market, but in comparison with orifice plates, nozzles, and venturis, there is little literature available, except from the manufacturers, relative to their C values. Flow tubes should be considered when low head loss is a critical factor.

Some manufacturers of venturi meters are incorporating the low head loss characteristics of flow tubes in their design so that the distinction between these types of devices is becoming less clear.

Equation (8.7.1) can be used for flow tubes as well as nozzles and orifices.

8.9.   ELBOW METERS

When pressure taps are placed on the inside and outside of an elbow, there will be a difference in pressure between the taps that varies as the flow squared. This provides the opportunity to determine flow. This setup is called an elbow meter and has the advantage of being inexpensive and producing no head loss in excess of what would normally occur in the bend. The primary disadvantage

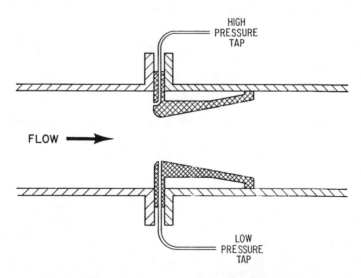

Fig. 8.8 Flow tube. [Reprinted with permission of J.P. Fath.]

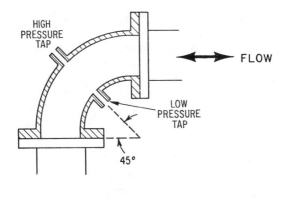

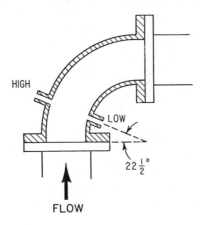

Fig. 8.9. Elbow taps. [Reprinted with permission of J.P. Fath.]

of the elbow meter is that unless it is calibrated in place, it is only accurate to 5%. This is because the results depend highly on the inside curvature of the bend, which varies from one manufacturer to another and even between two bends cast at the same time. The results are also very sensitive to the alignment of the manometer taps, which must be perpendicular to the interior elbow surface. Another disadvantage is that the elbow must be preceded by up to 25 diameters of straight pipe. Elbow meters are extremely inexpensive, since they require only tapping of the elbow at two places and installation of a differential pressure measuring device. The taps and measuring device should be constructed so as to avoid interference with pipe restraints.

Two elbow meters are shown in Figure 8.9. The top device has the advantage of operating equally as well when the flow is in either direction, since the taps are midway in the elbow. The second produces more stable results for flow in one direction but is fairly unreliable for flow in the other direction.

Because the head difference in an elbow meter is not exactly proportional to the flow squared, the exponent will vary depending on the size of the bend. Replogle, Myers, and Brust (1966) suggest using an expression of the form

$$Q = Mh^E \qquad (8.9.1)$$

where

$$Q = \text{flow, cfs}$$
$$h = \text{head difference, ft}$$
$$M, E = \text{coefficients.}$$

Instead of being a constant 0.5, E was found to vary with diameters from 3 in. to 12 in. according to

$$E = 0.489 + 0.0377D \qquad (8.9.2)$$

where $D =$ diameter, ft. Similarly, M was found to vary with the diameter and average elbow radius of curvature according to

$$M = (0.92 + 1.13D) \frac{\pi D^2}{4} \sqrt{\frac{2gr}{2D}} \qquad (8.9.3)$$

where $r =$ average radius of curvature of elbow, ft.

The above empirical relationships are based only on diameters from 3 in. to 12 in. and depend fairly strongly on the shape of the bend and the pipe material.

In general, it is best to calibrate an elbow in place using a pitot tube, although if identical conditions can be provided, the elbow can be calibrated more precisely in a laboratory.

8.10. PROPELLER, TURBINE, AND PADDLEWHEEL METERS

Water flowing through a pipe can turn propellers, turbines, and paddlewheels with a velocity which can be related to the velocity of the water. A number of devices have been developed based on this principle. Generally, they consist of some type of turbine and a counter which records the number of turns and can even convert that value directly into flow rate for a particular pipe and device. These devices range in style from permanent meters for large customers to portable devices which can be inserted into a pipe through a corporation stop.

The selection of a permanent customer meter requires an analysis of the difference between low flow and peak flow. If a significant amount of water use occurs at the low flow rates and accuracy is important, then a compound meter is required. If peak flow and typical low flow are fairly close (say 5:1) then a turbine meter may suffice.

Several portable units which can be inserted into the pipe through a corporation stop have recently made this type of meter attractive for field testing. These devices usually require a special electronic unit to process the signal produced to give the flow rate. Most require a power source, although some can use power from the flow itself.

While permanent turbine meters usually come equipped with straightening vanes and screens, field units, because of their very nature, cannot, and are subject to problems caused by unusual velocity patterns, misalignment, foreign material catching on the device, and inaccuracies caused by uncertainty of actual location of the device within the pipe. Turbine devices are fairly inexpensive and can be inserted into the pipe under pressure, but the associated electronic equipment can be quite expensive. Because the signal from these devices can be digitized and recorded, they have significant potential for use in water audits.

8.11. VORTEX SHEDDING METERS

Flow past a nonstreamlined object results in formation of vortices on alternate sides of the object. The rate of formation of the vortices is directly proportional to the velocity. This principle has been used as a basis for flowmeters which determine flow by counting the rate at which vortices are shed. This principle is similar to detecting wind speed by watching how fast a flag flutters.

Vortex detection can be accomplished using strain gages, thermal sensors, or magnetic detectors. The shape of the vortex shedding object and type of detector varies among manufacturers. Some vortex shedding devices can be inserted into a pipe under pressure, much like a pitot tube.

Usually, vortex meters are permanently mounted in the pipe and create an obstruction with some head loss. At present they are only competetive with other devices over a limited range of sizes. They require as much as 40 diameters of straight upstream piping and will not function at very low velocities. However, they have good accuracy and operate over a wider range of flows than many other devices.

8.12. ULTRASONIC METERS

Sound velocity in a liquid is affected by the velocity of that liquid. Thus, by measuring the velocity of sound traversing the flow, it is possible to calculate the velocity of the fluid. Several of these ultrasonic devices are available. They can be inserted through the pipe wall or merely clamped on the outside of the pipe so that there is no disturbance of the flow. The clamp-on units are especially attractive in that they can be moved around the system easily for such purposes as water audits. Typical ultrasonic meters are shown in Fig. 8.10.

The units generally require a 110 volt ac power source, and this somewhat limits their portability. Readings taken by the clamp-on units are sensitive to wall thickness, so there can be considerable error if the original wall thickness of the pipe is unknown or tuberculation or scale has formed in the pipe. Different transducers are required for each pipe thickness and pipe material. Thus costs can be fairly high if the meter is to be used for a variety of pipes. The

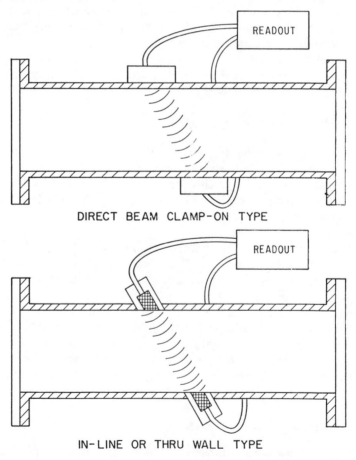

Fig. 8.10. Ultrasonic flowmeters. [Reprinted with permission of J.P. Fath.]

largest cost item is the main indicating and recording unit (essentially a small computer), which can cost several thousand dollars. Since concrete does not transmit sound well, the clamp-on units cannot be used for concrete pipe.

Special transducers exist which can be used to determine wall thickness of pipes. This can provide information for the ultrasonic meter.

Another type of device which operates on a similar principle is the *Doppler meter*. Rather than bouncing the sound waves off the opposite wall of the pipe, these units rely on the reflection of the waves from particles or bubbles in the fluid. These units do not work well for clear water, but are quite useful for aerated or dirty liquids.

8.13. ELECTROMAGNETIC FLOWMETERS

Flow of a conducting fluid through an electromagnetic field will generate a voltage related to the velocity of the fluid. In an electromagnetic flowmeter, a section of the pipe is replaced by the electromagnet with a voltmeter attached. The meter can be calibrated to read in any convenient flow units.

Electromagnetic flowmeters are quite accurate over a wide range of flows and cause no head loss. However, they are available only for permanent installations, and require a power source. Their primary disadvantage is their relatively high cost.

8.14. PITOT TUBES

The pitot tube was developed by Henri Pitot in 1732 to measure velocity in streams. The height of water in a column attached to a pitot tube is equal to the total head (also called the *stagnation pressure*) and can be given by

$$\text{Total head} = z + \frac{p}{\gamma} + \frac{v^2}{2g} \tag{8.14.1}$$

where

$$z = \text{elevation above datum}, L$$
$$p = \text{pressure}, M/LT^2$$
$$v = \text{velocity}, L/T.$$

To determine velocity, it is necessary to determine the static head ($z + p/\gamma$) and subtract it from the total head. The static pressure can be determined separately from the total head and then subtracted, but because the total and static heads are both large and their differences are small, it is much more accurate to determine the differences using a differential pressure gage. This difference can be related to velocity by

$$v = C\sqrt{2g\,\Delta h} \qquad (8.14.2)$$

where Δh = difference between static and total head, L.

A device which senses both total and static head is called a *pitot–static gage*. Some are made of a pitot tube which is inserted into the flow with a static pressure tap in the pipe wall, while others contain the static pressure tap as part of the tube. Cole (1935) developed a device in 1896 which contains two taps facing in opposite directions. This device gives fairly high values of Δh for a given velocity. What is more important is that this device was constructed with a packing gland such that it could be attached to a 1-in. corporation cock and inserted into a water main under pressure. This device, is known as the *Cole pitometer* $\text{\textcircled{R}}$, is the standard flow measuring device used in the water industry. It is shown with another pitot tube manufactured by Polcon in Fig. 8.11. Fig. 8.12 shows a pitot tube inserted into a clear pipe.

Pitot tubes (also referred to as *pitot rods*) are extremely popular for field testing and temporary installations because they are inexpensive, require no power, can be installed in a pipe under pressure, will operate in any size and type of pipe, are rugged, can sense flow in either direction, result in a small head loss only during the actual test, and for a given tip do not need calibration. The overall accuracy of the pitot tube and differential pressure device is on the order of 5%. The pitot tube is used for water audits, velocity measurements required as part of C factor tests, and testing large water meters in place.

Unlike most of the other meters discussed above (which measure flow), the pitot tube measures velocity at a point in the fluid. It is, therefore, necessary to integrate the velocity profile in order to calculate flow given velocity measurements at individual points. Given n velocities, the flow can be calculated from

$$Q = \sum_{i=1}^{n} v_i A_i \qquad (8.14.3)$$

where

$\qquad n$ = number of observations
$\qquad v_i$ = velocity at ith point, L/T
$\qquad A_i$ = area of annular ring for which velocity v_i is accurate, L^2.

The cross-sectional area of the pipe is divided into annular areas rather than other shapes (say horizontal slices) because velocity varies more in the radial direction than other directions.

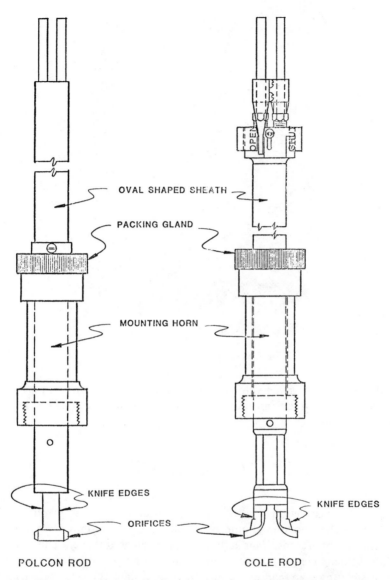

OVAL SHAPED SHEATH

PACKING GLAND

MOUNTING HORN

KNIFE EDGES

KNIFE EDGES

ORIFICES

POLCON ROD

COLE ROD

Fig. 8.11. Pitot tubes used in water distribution applications. [Reprinted with permission of C.F. Buettner and G.B. Cole.]

Table 8.2 shows a suggested method for recording the results of the pitot traverse. The units on differential pressure are left blank because they depend on the device used. Eq. (8.14.2) requires the differential head to be units of feet for velocity in feet/second, since the coefficient C is dimensionless. This conversion must be made before calculating velocity.

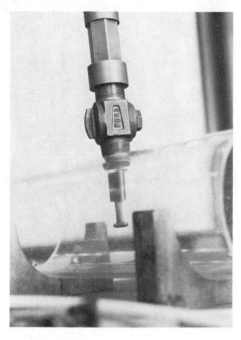

Fig. 8.12. Pitot tube inserted into pipe.

Calculating flow, given velocities, is considerably easier if the pipe cross section can be divided into equal-sized areas so that Eq. (8.14.3) can be simplified to

$$Q = (A/n) \sum_{i=1}^{n} v_i \qquad (8.14.4)$$

where A = cross section area of pipe, L^2. In order for this formula to be appropriate the velocities should be measured at the midpoints of annular rings having equal areas. Usually 10 velocity readings are taken. This requires making velocity measurements at the 10 points shown in Fig. 8.13. Since it is not possible to make readings at exactly those points, the velocity is usually determined at more than 10 points from top to bottom (with the points concentrated away from the center) and a graph of velocity versus dimensionless distance from bottom of pipe is plotted. An example of a sheet used to plot the profile is shown as Fig. 8.14. The velocity at each horizontal line can be read and averaged to give the average velocity which can be multiplied by the area to give flow. In these calculations the actual rather than the nominal pipe area should be used.

When the velocity can be reasonably approximated by the power law

Table 8.2.   Pitot Tube Results.

Location:

Date: Time:

Conducted by:

Pitot Tube Coefficient:

Pipe Internal Diameter: in.

Differential Pressure Device: manometer (SG of indicator)
            (circle one)   differential gage
                              transducer

Distance from Bottom (in.)	DP Device Reading ( )	Velocity (ft/sec)

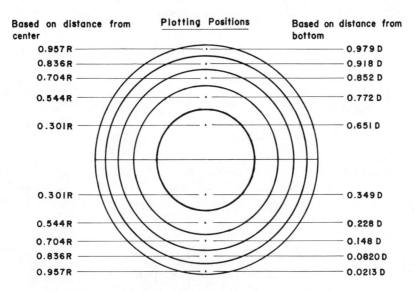

Based on distance from center	Plotting Positions	Based on distance from bottom
0.957 R		0.979 D
0.836 R		0.918 D
0.704 R		0.852 D
0.544 R		0.772 D
0.301 R		0.651 D
0.301 R		0.349 D
0.544 R		0.228 D
0.704 R		0.148 D
0.836 R		0.0820 D
0.957 R		0.0213 D

Fig. 8.13. Location of plotting points for ten-point pitot tube traverse.

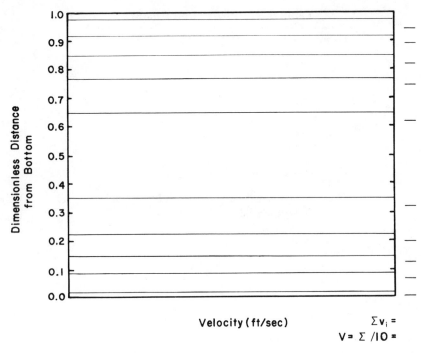

Fig. 8.14. Form for calculating average velocity from pitot tube traverse.

formula, the flow can be determined by determining the constants in the power law formula and integrating analytically:

$$v(r) = a\,[(R - r)/R]^{b} \qquad (8.14.5)$$

where

$v(r)$ = velocity r feet from the center of the pipe, ft/sec
R = pipe radius, ft
a, b = constants.

The constants a and b can be determined by plotting v versus $(R - r)$ on log-log graph paper. The constant a will be the value of v at the center of the pipe, i.e., $(R - r)/R = 1$, and b is the slope, which can be determined between any two points on the line as

$$b = \frac{\log\,(v_2/v_1)}{\log\,[(R - r_2)/(R - r_1)]}. \qquad (8.14.6)$$

By analytically integrating Eq. (8.14.5), Q can be shown to be

$$Q = 2\pi a R^{b+2} \left[\frac{1}{b+1} - \frac{1}{b+2} \right] \qquad (8.14.7)$$

The pitot tube occupies some space in the pipe and, as such, it tends to cause an increase in velocity, especially when it is fully extended into the pipe. Buettner (1980) calculated the area taken up by the Polcon pitot tube when it is inserted such that the tip is at the midpoint of the pipe. This area can be related to the diameter of the pipe by

$$dA = 0.0128 D^{1.14} \qquad (8.14.8)$$

where
$$dA = \text{area occupied by rod, ft}^2$$
$$D = \text{diameter, ft.}$$

Readings taken at the opposite end of the pipe from the tap will be affected more than readings taken near where the tube was inserted, but this average value is adequate. The correction factor is 2.7 percent of the area for a 6-in. pipe, but decreases to less than one-half percent for a 48-in. pipe.

Once velocity profiles have been determined for several flow rates for a given pipe, it is possible to develop a simple relationship for determining flow based on the velocity at the center of the pipe. The ratio of the average velocity to velocity at the centerline of the pipe is known as the *pipe factor*; it varies from roughly 0.75 to 0.97 for pipes with typical velocity profiles. The value may even exceed 1 for a pipe with, for example, a partially closed gate valve upstream of the test section. Once the pipe factor is determined it can be used to calculate flow rate when only the velocity at the centerline is know, since the pipe factor has been shown to be fairly independent of flow rate in a given pipe. Folsom and Iverson (1949) found the pipe factor to range from 0.89 in smooth pipe at high Reynolds numbers to 0.78 for rough pipe ($\epsilon/D = 0.012$) at low Reynolds numbers. Significantly different pipe factors can be found near bends and fittings. The flow can be given by

$$Q = A \, (\text{PF}) v \, (0) \qquad (8.14.9)$$

where
$$v \, (0) = \text{velocity at centerline, } L/T$$
$$\text{PF} = \text{pipe factor}$$
$$= v \, (0)/V.$$

Another method used to avoid making a complete traverse of the pipe with the pitot tube is to position the tip at a point at which the velocity is equal to the average velocity. This is generally about 30% of the pipe radius from the pipe wall. The exact distance depends on pipe roughness and Reynolds number, and considerable error can result from the use of typical values if there is any kind of obstruction or bend upstream which effects the velocity profile. For this reason, use of the pipe factor is considered more accurate for determining average velocity.

EXAMPLE. A pitot tube with a coefficient of 0.83 was used to measure velocity in a 24-in. internal diameter pipe. The differential pressure readings were made with an air-filled manometer and the line pressure was 55 psi. The manometer reading and the distance from the bottom of the pipe are given in the table below in columns 1 and 2.

The velocity at each measuring point can be given by

$$v = 0.83 \sqrt{2(32.2)(1 - 0.0045)\Delta h}$$

where Δh is the manometer reading in feet and 0.0045 is the specific gravity of air at 55 psi as given by Eq. (8.3.4). The velocities calculated by this formula are shown in column 3 of the table and are plotted in Fig. 8.15.

Distance from Bottom (in.)	Manometer Reading (in.)	Velocity (ft/sec)
1.2	9.1	5.8
2.4	11.1	6.4
4.8	13.5	7.1
7.2	15.1	7.5
9.6	16.4	7.8
12.0	17.3	8.0
14.4	16.3	7.7
16.8	15.0	7.4
19.2	13.4	7.0
21.6	11.1	6.4
22.8	9.1	5.8

The calculations to obtain the average velocity of 6.7 ft/sec are shown on the figure. The pipe factor is 0.83 = (6.7/8.0). (It is only a coincidence that the pipe factor and the pitot tube coefficient are both 0.83 in this example.)

While the pitot tubes accurately measure dynamic pressure over a wide range of conditions, measurement of static pressure depends somewhat on the Reynolds number and pipe geometry. This means that the constant C in Eq. (8.14.2) is not exactly constant and depends on the conditions for which the tip was calibrated. Cole (1935) reported that the pitot tube coefficient varied

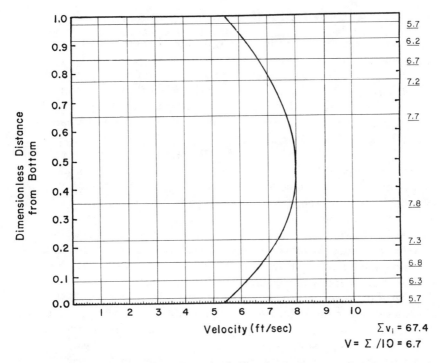

Fig. 8.15. Example of pitot tube traverse.

from 0.89 to 0.86 as velocity increased from 2 to 10 ft/sec. Hauserman (1980) reported that the coefficient varies from 0.905 to 0.854 as velocity varied from 0.3 to 20 ft/sec. The effect of velocity variation is generally small, but it does limit the claimed accuracy of the device to ±5%, even though in most cases it is considerably more accurate.

Pitot tubes should always be preceded by at least 50 pipe diameters of straight pipe so that the velocity profile is not influenced by secondary currents. If this cannot be done, the error can be reduced by making traverses in more than one direction. For example make one from top to bottom and a second from side to side (Robertson and Clark, 1977).

To eliminate the need for making a complete traverse of the pipe with a pitot tube, several manufacturers have developed what are called *distributed pitot tubes*. These devices consist of a rod containing several upstream holes located at such locations that the average of the readings is the average velocity. This method is generally more accurate than a pitot traverse, but a given rod can only be used on a single size of pipe internal diameter. Thus, for field testing, the engineer must have a different rod for every internal diameter that will be encountered. This can be fairly expensive, although less expensive than a set of ultrasonic transducers.

8.15.   MEASURING DISCHARGE

Engineers are not only interested in knowing flows in pipes, but also discharge from pipes, hydrants, and hoses. One way to measure such discharges is to connect a section of pipe to the opening and use one of the devices described in the previous sections. Another technique is to allow the water to run into an open channel where the flow can be measured using a weir or flume. These methods are, however, unnecessary, since there are special techniques available for measuring discharge.

The simplest method for measuring discharge is the "bucket and stopwatch" technique. This involves measuring the length of time it takes for a container of known volume to fill. The bucket can also be placed on a scale and the increase in weight with time can be noted and converted into flow units. The bucket should be large enough that it takes at least 10 seconds to fill, and the results will be most accurate if the time to fill the bucket is on the order of several minutes. In most water systems, the discharges of interest are so large that this approach is useful only for special applications, such as when a hydrant is discharging into a sump with known volume.

When water is discharged horizontally from an orifice, the stream will curve downward under the influence of gravity until it strikes the ground. The horizontal component of the velocity does not change significantly during that time, so it is possible to calculate the velocity based on the distance the water travels and the height of the outlet. This method, referred to as the *trajectory method*, is shown in Fig. 8.16. The velocity determined this way is that at the narrowest point of the jet, called the *vena contracta*. This velocity is given by

$$v = \frac{x}{\sqrt{2y/g}} \tag{8.15.1}$$

where
g = acceleration due to gravity, L/T^2
x,y = horizontal and vertical components of distance from vena contracta to impact with ground, L
v = velocity at vena contracta, L/T.

The discharge can be determined by multiplying the area of the vena contracta by the velocity. This method is fairly inaccurate because of the inevitable difficulty in measuring x, y, and the area of the vena contracta. Two variations on this method, the *California method* and the *Purdue method*, are applicable to discharge from pipes (Guthrie, 1981).

A fairly easy method to determine discharge from outlets is to insert a pitot

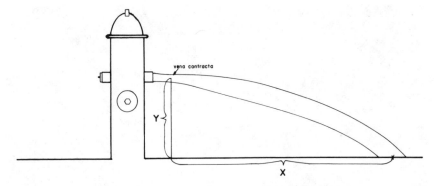

Fig. 8.16. Trajectory method for determining discharge.

gage into the flow. Since virtually all of the head is in the form of velocity head, it is easy to calculate the velocity from the pressure measurement. The flow can be determined by

$$Q = 29.8 D^2 C \sqrt{p}$$ (8.15.2)

where

Q = discharge, gpm
D = outlet diameter, in.
p = pressure detected by pitot gage, psi
C = coefficient.

The coefficient C is usually taken to be 0.90 for a hydrant when flow occurs across the entire outlet. In general, the $2\frac{1}{2}$-in. outlet should be used for the test because its use results in more accurate readings for the same flow rate than the large pumper outlets. If the pumper outlets are used the discharge should be multipled by 0.83 for pitot gage readings in excess of 7 psi. The coefficient increases to 0.84, 0.89, and 0.97 as the pressure decreases from 6 to 4 to 2 psi.

Some pitot gages have square root scales so that the flow can be read directly from the gage. An example is shown in Fig. 8.17. Pitot gages are usually hand held, although as shown in Fig. 8.17 some are available which can be clamped into place on a $2\frac{1}{2}$-in. outlet.

Portable hydrant flow meters, essentially turbine flowmeters which can be mounted on a hydrant outlet, are also available. These devices are primarily used for keeping track of water use at temporary connections during construction projects, but can be used for flow testing. Since these gages record the total volume used, two readings taken over a known period of time must be divided by the elapsed time between readings.

Fig. 8.17. Pitot gage for measuring hydrant discharge.

In conducting any kind of discharge measurement the engineer must take care to avoid causing any flooding or erosion problems from the discharge from the hydrant.

8.16. DISTANCE

Distance measurements are required in making hydraulic gradient calculations for head loss tests and in describing the pipe network for a computer model. Whenever possible the distances should be measured with a tape, measuring wheel, or surveying equipment. For modeling studies distances must be measured from maps and aerial photographs. Straight line distances can be read from maps using an engineer's scale, while curved lines are best taken from maps with a pocket size measuring wheel, as shown in Fig. 8.18.

The key to accurate measurement of pipe lengths is to measure along the actual path the pipe takes, as pipes do not always lie in the middle of the street. In conducting head loss or C factor tests where head loss is measured between two hydrants or other points at which pressure can be measured, the distance used must be the distance between the points where the hydrant lateral branches off the main, rather than the distance between the hydrants. Fig. 8.19 shows how distance should be measured in conducting a C factor test when

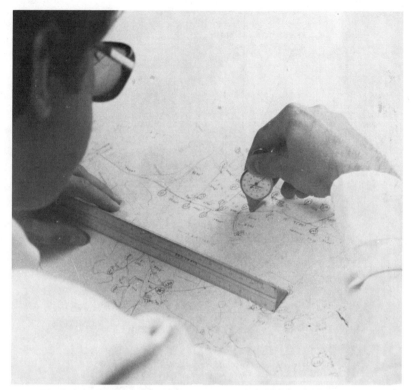

Fig. 8.18. Measuring pipe length with measuring wheel for model input.

head is measured by the parallel pipe method, which is discussed in greater detail later.

8.17. PIPE DIAMETER

The internal diameter of a pipe is a necessary input to most hydraulic calculations, and for new concrete, asbestos cement, and most steel pipe, it is identical with the nominal diameter. However, for new ductile iron and some plastic pipe, for which the outer diameter was made to match the outer diameter of old fashioned cast iron pipe, the internal diameter is larger than the nominal diameter, since wall thickness is smaller. In old pipe the internal diameter may have changed considerably as a result of tuberculation or scaling in the pipe.

The only way to precisely know the internal diameter of pipe is to cut out a section and measure it or use a pipe caliper. A pipe caliper (Fig. 8.20) can be

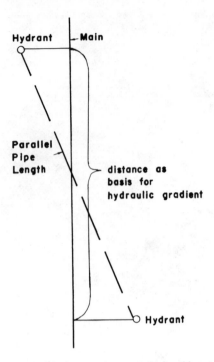

Fig. 8.19. Distance measurement for head loss test.

inserted into a pipe through a corporation cock exactly as a pitot tube is. The caliper is lowered into the pipe until the extreme point of the hook-shaped rod (x) reaches the opposite wall; that point is then marked on the rod outside the pipe. The hook is then rotated 180 degrees and the rod is pulled upward until the tip of the hook (y) reaches the top of the pipe. The difference between these two distances plus the vertical dimension of the hook (z) is equal to the internal diameter of the pipe.

The internal diameter of new pipe can be found in the manufacturer's specifications for that pipe. The internal diameter of newly cleaned and lined pipe is usually close to the original diameter.

If confronted with an exposed pipe of unknown diameter, the engineer can at least gain an approximate feel for the nominal size by wrapping a tape measure around the pipe and calculating the diameter as $D = C/\pi$, where C is the circumference.

The question of whether to base the Hazen-Williams C factor on the nominal or actual internal diameter can have serious implications, especially when payment for cleaning and lining jobs is based upon the C factor after the

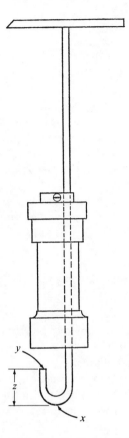

Fig. 8.20. Pipe caliper. [Reprinted with permission of C.F. Buettner.]

job or the improvement in C. Another instance when the difference is important is when C is used in a model of the pipe network. If the nominal diameters
are used as input to the model but the C factors are calculated based on actual
diameter, the model will not predict head loss correctly.

EXAMPLE. A head loss test for a nominal 10-in.-diameter (actual 10.5-in.-diameter)
pipe indicates that 1.11 ft of head loss occurs in 300 ft of pipe at a flow of 800 gpm.
What is the C factor based on nominal and actual diameter?
   The C factor based on nominal diameter is

$$C = \left[\frac{(10.4)(300)}{10^{4.87}(1.11)} \right]^{0.54} (800) = 136.6$$

In contrast, based on actual diameter it is

$$C = \left[\frac{(10.4)(300)}{10.5^{4.87}(1.11)} \right]^{0.54} (800) = 120.$$

Which diameter in the preceding example is correct? It actually depends on how the values for C are to be used. If the values are to be used in a computer model where the model user only knows nominal diameters for most pipes, then the C based on nominal diameter should be used. If it is to make up a table of C factors such as the one in Chapter 2, the actual diameter should be used.

The error caused by mixing diameters can be significant. Suppose using the above example, a C of 120 based on actual diameters was used with the nominal diameter of 10 in. to predict head loss in the pipe at 800 gpm. The answer would be

$$h = \frac{(10.4)(300)}{10^{4.87}} \left(\frac{800}{120} \right)^{1.85} = 1.40 \text{ ft.}$$

Yet the head loss in that pipe is known to be 1.11 ft at 800 gpm—an error of 27%. The actual diameter in this problem is based on thin-wall (class 50) ductile iron pipe with a small diameter, a situation that would yield a large error. A similar error, in the opposite direction, can occur for severely tuberculated pipes. Therefore, when using a C factor for a given pipe, the engineer should know the diameter on which it is based.

8.18. ELEVATION

Accurate elevation data are important for pipe network models, hydraulic gradient studies, and measuring head losses with a two-gage method. Ideally, elevation data will not be the weak link in a model study but when data are only available on topographic maps with 20-ft contour lines, the confidence that can be placed in the model's pressure predictions will be limited primarily by elevation data.

There are other sources of elevation data for model calibration. For example, gas utilities sometimes have better elevation data than water utilities, or plans from a recently completed sewer project may contain very good elevation data. In some cases, aerial photographs with 2- or 5-ft contour lines are available. Water utilities may want to consider joining together with gas and power utilities and local planning commissions in developing contour maps with contour intervals which will not be the limiting factor in hydraulic analyses.

If only a few elevations are required to determine the elevation of the hydraulic grade line in an area to locate pressure problems or to be used for precise model calibration, it may be worthwhile to survey in those points from known benchmarks. If this is the case, it is important to use the same datum for all elevations.

In measuring head loss using two gages, the elevations must be known to at least the nearest foot. This can be done using surveying equipment or the two pressure gages. To measure difference in elevation using two pressure gages, the flow in the pipe between the two gages must be stopped, but there must not be any valves closed between the two pipes. The difference in elevation will be the difference in pressure gage readings converted to elevation units. This method is only workable for those rare occasions where it is possible to completely shut off flow in a pipe during testing.

8.19. CALCULATIONS USING TEST RESULTS

The preceding sections described how to measure quantities such as velocity in pipes and pressure differentials. While these quantities are important in themselves, they are also used to calculate other quantities such as pipe roughness, fire flow delivered from a set of hydrants, or accuracy of meters.

Each of these test procedures is described in the following sections, and formulas needed to calculate the results are given. The reader should refer to the previous sections for discussion of the techniques used in making the required measurements.

8.20. PIPE ROUGHNESS AND C FACTOR

Tests to determine pipe roughness (or C factor) are conducted in conjunction with model studies or when utilities are deciding whether to clean and line water mains. These tests are conducted on a length of pipe along which flow is virtually constant (i.e., there may be some domestic customers but no open interconnections with other mains) and pressure can be measured at both ends of the pipe. The pipe should be of one diameter and type, as it is difficult to develop roughness values when pipe characteristics change along the test section.

The parameters which must be measured to calculate pipe roughness are: length, diameter, head loss (or pressure and elevation differences), velocity (or flow), and water temperature (for calculating kinematic viscosity for Reynolds number).

The length of the test section depends on the method for measuring head loss. If a parallel pipe with no flow and a differential pressure gage is used, the length should be several hundred feet. For example, if a 5 psi differential gage

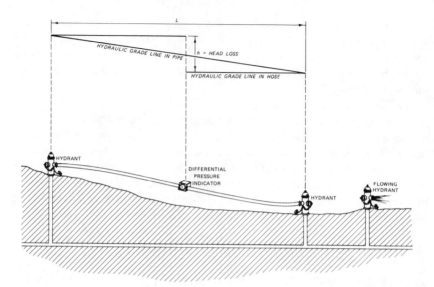

Fig. 8.21. Measuring head loss by parallel pipe method.

is used, the length should be sufficient to yield at least a 2 psi head loss based on expected flow and roughness. If the head loss is measured using two pressure gages (a far less accurate method), a considerably longer test section is needed, as pressure gages are seldom accurate to better than 1 psi.

The parallel pipe method for measuring head loss, illustrated in Fig. 8.21, is based on the fact that head is constant in a pipe with no flow. Therefore the head seen by each side of the differential pressure gage is exactly the head at each end of the test section. There is also no need to measure hydrant elevations when the parallel pipe is used. Usually, the parallel pipe is a hose and it is desirable to use pressure snubbers on each side of the differential gage to dampen out fluctuations due to water hammer. Otherwise, reading the gage will be difficult. It is critical that there be no leakage from the parallel pipe. A leak creates a flow in the parallel pipe which induces head loss and thus renders the readings inaccurate.

As stated in the previous section on pipe diameter, it is essential to state which diameter is used to calculate roughness if the nominal and actual diameters differ. The decision as to which value to use depends on the ultimate use of the roughness values. A similar question arises concerning minor losses in the pipe. If the roughness in the pipe only is desired, then the losses occurring due to bends and valves should be explicitly accounted for using minor loss coefficients, but if the effective roughness of the pipe, fittings and bends is to be lumped together, then the actual pipe length should be used. The decision depends on the ultimate use of the data. If they are to be used in a pipe network model, will the model user ignore minor losses or will the minor losses be accounted for as equivalent pipe? The important thing is to be consistent, regardless of the approach used. The error involved in ignoring or double counting minor losses is, however, small for most distribution systems.

The velocity in the pipe should be determined using a pitot tube or some other main metering device such as an ultrasonic meter or turbine meter which can be inserted into the pipe. Ideally, the velocity should be measured at each end of the test section so that any water use or inflow along the line can be detected. Unfortunately, this is usually quite expensive if excavation to the top of the pipe is required. Once a traverse of the pipe has been made so that a pipe factor can be calculated the engineer should change the flow (for example, by opening one or two hydrants downstream of the pipe) and measure the velocity and head loss for that condition. In this way, it is possible to obtain roughnesses or C factors over a range of flow rates. This is quite helpful since the C factor in rough pipes varies slightly with the velocity. The velocity and head loss will change during the course of the test, especially for small pipes, due to fluctuations in water use. The results of the test should therefore be a set of pairs of velocity and head loss measurements.

Once the tests are run, the C factor can be calculated as follows:

$$C = \frac{8.71\,V}{D^{0.63}\,(h/L)^{0.54}} \qquad (8.20.1)$$

where

C = Hazen–Williams C factor
V = velocity, ft/sec
D = diameter, in.
h = head loss in pipe length L, ft
L = length of test section, ft.

In order to determine the pipe roughness, the friction factor must first be calculated from the Darcy–Weisbach equation:

$$f = \frac{5.37hD}{LV^2} \qquad (8.20.2)$$

where f = friction factor. Once the friction factor is known it is necessary to know the Reynolds number to obtain pipe roughness. The kinematic viscosity, which depends on temperature, must be determined first. Table 8.3 gives kinematic viscosity as a function of temperature. The Reynolds number can then be calculated as

Table 8.3. Kinematic Viscosity of Water

Temperature (deg F)	Kinematic Viscosity 10^{-5} (ft^2/sec)
32	1.91
40	1.66
50	1.41
60	1.22
70	1.06
80	0.93
90	0.83
100	0.74

$$N_r = \frac{VD}{12\nu} \qquad (8.20.3)$$

where ν = kinematic viscosity, ft^2/sec.

Knowing the friction factor and the Reynolds number, the engineer can determine relative roughness of the pipe from the Moody diagram or the Colebrook–White equation solved for roughness:

$$\frac{\epsilon}{D} = 10^{(1.14 - 1/\sqrt{f})/2} - \frac{9.35}{N_r \sqrt{f}} . \qquad (8.20.4)$$

Note that when this formula is applied to smooth, oversized pipes with f based on nominal diameter, the roughness can appear to be negative.

In reporting the results of head loss or pipe roughness tests, it is important to report not only the roughness or C, but also the velocity, diameter, and pipe length on which it is based. This will alert the eventual user of the data to problems which could occur at different velocities or diameters, and will enable the user to generate alternative values based on specific needs.

Table 8.4 gives a suggested form for recording the results of tests to measure pipe roughness and C. It is based on determining head loss by the parallel pipe method and velocity by a pitot tube, although it could be used for other techniques with only minor modifications.

8.21. INDIRECT METHODS OF DETERMINING ROUGHNESS

The methods for determining pipe roughness described in the previous section yield fairly accurate results but are fairly costly to run. The major costs are usually associated with excavation to the top of the pipe, trench shoring, pipe tapping, backfill, and repaving. These costs probably amount to three times

Table 8.4.   Pipe Roughness Test Results.

Location:

Date: Time:

Conducted by:

Pipe Diameter: in. (Nominal)

                        in. (Actual)

Length of Test Section: ft (Actual)

                        ft (Equivalent w/minor losses)

Water Temperature: deg F*

Kinematic Viscosity: ft^2/sec*

Describe Head Loss Measuring Technique:

Pipe Factor for Pitot Tube:
(Record traverse data on pitot tube data sheet)

	Test 1	Test 2	Test 3	Test 4
Head Loss (ft)				
Pitot Reading				
Velocity (ft/sec)				
C				
f*				
Reynolds Number*				
ϵ/D*				

Calculations based on   diameter and   length.

*Not required if only C is to be calculated.

the cost of conducting the test on an above-ground pipeline. Without tapping the pipe, there is no way to directly measure velocity. In some cases a pipe can be isolated such that all of the flow in the pipe is discharged through a hydrant or blowoff where it can be measured. There is also a way to calculate flow and C factor (and hence roughness) if head loss can be measured for several flow rates while the difference in flow rate is observed.

Tests can be conducted when a pipe is isolated from the rest of the system so that water can reach the pipe only from one direction. A hydrant is then opened at the downstream end of the pipe and the flow from the hydrant is the flow in the pipe. It is a straightforward matter to then calculate the C factor using the following equation

$$C = \frac{3.54Q}{D^{2.63}(h/L)^{0.54}} \qquad (8.21.1)$$

where

C = Hazen–Williams C factor
Q = flow in pipe, gpm
D = diameter, in.
h = head loss, ft
L = length over which head loss occurs, ft.

Eq. (8.21.1) is easy to use when there are no water users downstream of the pipe (i.e., $h = 0$ when the hydrant is closed). The problem is that it is difficult to isolate mains in a distribution system to the extent required to use Eq. (8.21.1).

Another method for measuring C in large mains involves isolating the portion of the distribution system served by the main as shown in Fig. 8.22. In this way an unknown flow Q is allowed to enter the downstream service area. The key is that there must be no other way for water to enter that portion of the service area during the test (i.e., valves A, C, and D must be shut). (The previous method requires that valve B be shut.) Selecting the method with which to isolate the pipe depends on (1) the ease with which the valves can be shut, and (2) the need to minimize disruption of service during the test. This method, which involves isolating an entire service area, has the added advantage of yielding an estimate of water use in the service area during the test.

The pipe length and diameter can be determined by one of the methods described earlier, and the head loss is measured when no hydrants are open (h_1). This corresponds to some unknown flow rate (Q_1), as shown below:

$$\frac{h_1}{L} = \frac{10.4}{D^{4.87}}\left(\frac{Q_1}{C}\right)^{1.85} \qquad (8.21.2)$$

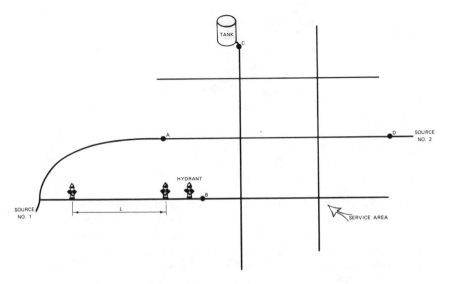

Fig. 8.22. Isolating main for head loss test.

where the subscript 1 refers to the quantity when the hydrant(s) are closed. When the hydrant(s) are discharging a flow Q_f, the head loss can be given by

$$\frac{h_2}{L} = \frac{10.4}{D^{4.87}} \left(\frac{Q_2 + Q_f}{C} \right)^{1.85} \tag{8.21.3}$$

where

Q_2 = water use in the service area when the hydrant is open, gpm
Q_f = discharge from hydrant, gpm.

The two preceding equations above contain three unknowns Q_1, Q_2, and C. The key to calculating C is realizing that Q_1 and Q_2 are fairly close to one another and their ratio can be given as

$$r = Q_2/Q_1. \tag{8.21.4}$$

In general the water use in the service area will not change because the hydrants are open, but the discharge from any orifice will decrease slightly because the pressure in the service area will drop slightly during the test. This effect is offset by the fact that some uses (e.g., filling toilet tanks) will continue longer during the time the hydrant is open. The upper limit on r is unity (use will not increase) while the lower limit, based on the fact that discharge depends on head to the one-half power, is $\sqrt{P_2/P_1}$, where P_2 is the pressure at some representative point when the hydrant is open and P_1 is the pressure at the same location when it is closed.

It is possible to solve the above equations simultaneously for C and water use to yield

$$C = \left(\frac{10.4L}{D^{4.87}h_2} \right)^{0.54} Q_f \left[\frac{1}{1 - r(h_1/h_2)^{0.54}} \right] \qquad (8.21.5)$$

and

$$Q_1 = \left(\frac{h_1 D^{4.87}}{10.4L} \right)^{0.54} C. \qquad (8.21.6)$$

The expression to the left of the brackets in Eq. (8.21.5) represents the C factor if all of the flow in the pipe is measured at the hydrant. The expression in the brackets is a correction to that term which takes into account the fact that the pipe carries a considerable amount of flow in addition to the discharge from the hydrant. When the term in the brackets is equal to one (no other water use), Eq. (8.21.5) reduces to Eq. (8.21.1). If the head loss when the hydrant is open is large ($h_2 \gg h_1$), then the effect of inaccuracies in r is minimized, although r would be slightly farther from one in that case. The value of r will approach one as the length of time the hydrant is open increases.

The accuracy of the test depends strongly on the fact that the discharge from the hydrants is of the same order of magnitude or larger than the flow going to the service area. Before going to the field to conduct such a test, the engineer should evaluate the flow required to produce measurable head loss in the pipe, since as h_2 approaches h_1 the term in brackets becomes very large and any small error in measurement will result in large errors in C. As a very rough rule-of-thumb, one hydrant can be opened for mains less than 12 in., two hydrants for mains between 12 in. and 24 in., and three for large mains, although as the size exceeds 30 in. the flooding problems resulting from conducting such a test may make it infeasible. The engineer should also be certain that the differential pressure measuring device for determining head loss has an adequate scale for the task.

EXAMPLE. Consider a 500-ft-long, 16-in. internal diameter pipe for which the head loss is 0.23 ft when no hydrants are open. When one, two, and three hydrants are opened downstream, the head losses measured are 0.85, 1.29, and 1.81 ft for hydrant discharges of 800, 1200, and 1600 gpm, respectively. Calculate C for each test, using 1.0 for r then 0.95.

Substituting into Eq. (8.21.5) and simplifying gives

$$C = \frac{0.0692 Q_f}{h_2^{0.54}} \left[\frac{1}{1 - r(0.23/h_2)^{0.54}} \right].$$

The results of calculations for C for all three observations and both values of r are tabulated below:

Q_f	h_2	$r = 1.0$	$r = 0.95$
800	0.85	114	119
1200	1.29	116	119
1600	1.81	115	119

These values correspond to Q_1 ranging from 745 to 778 gpm.

The example above indicates that there will be considerable uncertainty in determining C with the indirect method (on the order of 10%), but in some situations this level of accuracy is acceptable and the costs and inconvenience associated with other methods can be avoided.

8.22. HYDRAULIC GRADIENT TESTS

Finding the weak links in a water distribution system can be done easily by conducting a hydraulic gradient test (Hudson, 1954). In this test, the elevation of the hydraulic grade line (static head) is determined along a path from the source to some arbitrary ending point. At a number of points along the system located at intervals of from 500 ft to a mile, the pressure and the elevation at which that pressure was measured is recorded. The static head at that point is calculated using

$$H = z + 2.31p \tag{8.22.1}$$

where

H = static head, ft
z = elevation of pressure gage, ft
p = pressure, psi.

The static head is then plotted versus distance from the source.

On the plot of the hydraulic grade line, weak links in the system show up as very steep slopes, while for sections with adequate carrying capacity the plot will be relatively flat. For the test to be meaningful, it must be carried out during a time when the demand is roughly constant and settings on pumps and valves are not being changed. In general, it is best to run the test during high-demand periods, since that is when problems can be more easily detected. For example, a test should be run between 9 and 11 am or 1 and 3 pm, rather than between 6 and 8 am, when the demands tend to fluctuate.

Weak links in a system result from: (1) inadequate pipe sizes, (2) low carrying capacities due to roughness, or (3) closed or partially closed valves. The engineer should work backward through this list in trying to locate problems. Closed or damaged valves can be repaired or adjusted at considerably less cost then cleaning and lining the main, which in turn is generally less costly than replacing or paralleling a main. (Refer to Chapter 6 for details on this type of evaluation.) If excessive head loss is only a problem during high-flow periods, the problems can sometimes be corrected by installing elevated storage to reduce peak flows in major transmission lines. If a bad valve is suspected in an area, the static head can be measured at fairly small intervals (e.g., every hydrant) to locate on the problem.

The primary limitation of hydraulic gradient tests is that they are conducted along a single line in a system, but real systems radiate out in many directions. An overview of the system can be obtained by making a set of such graphs, or by plotting hydraulic grade contours on a map. Usually, if the problem is sufficiently complex, a hydraulic model of the system is required for the analysis.

The elevation and pressure data should be accurate to the nearest two feet for the hydraulic gradients to be meaningful. This rules out the possibility of picking elevations off contour maps with 20-ft intervals. If this type of test is to be run more than once, it may be worthwhile to simply survey in the elevations of hydrants and distances between hydrants where pressure is to be measured.

EXAMPLE. A complaint of low pressure has been made by a customer near the end of the system, so pressures are read along the main to that customer from the source and the elevations and distances between the test points are carefully measured. The distance from the source, elevation, and pressure are given in the first three columns of the table below. Determine which segment of the pipeline is most responsible for the problem by plotting the hydraulic gradient.

The static head and slope of the hydraulic grade line are given in columns 4 and 5 of the table, and the grade line is shown in Fig. 8.23.

Distance (ft)	Elevation (ft)	Pressure (psi)	Head (ft)	Gradient (/1000)
0	235	110	489	
1235	350	60	488	0.8
2480	312	75	485	2.4
4829	284	86	482	1.3
6410	325	67	479	1.9
8174	310	65	461	10.2
12891	280	76	455	1.3
15140	340	48	451	1.8

The problem in the system occurs between locations 6410 and 8174. The engineer

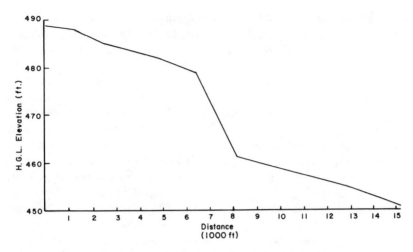

Fig. 8.23. Example of hydraulic gradient test results.

may want to conduct more tests in the area to better identify the problem. Another approach is to estimate velocities in the area and calculate the corresponding C-factor. If it is low, the main may be a candidate for cleaning and lining.

8.23. FIRE FLOW TESTS

Provision of water for fire fighting is an important consideration for virtually all water distribution systems. Evaluating the capability of the distribution system to provide water for fire fighting is difficult to do precisely, since this would involve connecting pumper engines to the system and creating exactly the kind of flows required during a fire. This type of simulation, while producing an accurate reading on the flow that can be delivered for fire fighting, creates problems with flooding, wasted water, and low pressure in the system for regular customers during the test.

The key to evaluating a distribution system is to be able to extrapolate the flow that could be delivered for fighting fires based on tests conducted at significantly lower flows and without pumper engines. Simply measuring the discharge from a hydrant and multiplying by the number of hydrants that can be used in a given area is inaccurate, since it does not take into account the hydraulics of the distribution system in getting water to the hydrants.

The most accurate way of predicting the flow that can be delivered for fire fighting is to use a properly calibrated pipe network model to simulate the flows. The keys to this approach are: (1) the model must be calibrated using fire flow test data as described in Chapters 2 and 4, and (2) the model of the system in the vicinity of the simulated fire flow must include virtually all pipes (i.e., the system near the hydrants should not be skeletonized as much as the remainder of the system). Under no situation should a computer model of the system be used to predict fire flows without first being calibrated using fire flow data for that area. There is no substitute for data collected during a test.

If a computer model is not used, the flow that can be delivered can be estimated using a formula developed by the Insurance Services Office (ISO, 1973) or a more general equation that will be presented below. Before deriving the equations, the procedure developed by ISO is presented. This is followed by some suggested modifications to make the results more useful to engineers who must evaluate the systems.

To conduct the test according to ISO procedures, a pressure gage is connected to a hydrant or some other point at which accurate pressure readings can be taken. This is called the *residual hydrant*. All pressure readings are taken at this hydrant, which should be further from the water source than the hydrants to be flowed, but on the same mains. Then one or more hydrants are opened and flowed, with the flow measured at each using a pitot gage as described in Section 8.15. The pressure at the residual hydrant is measured before the hydrants are opened (P_s) and when the hydrants are flowing (P_t). The sum of the flows from the hydrants (Q_t) is also recorded. Fig. 8.24 shows a fire flow test and head loss test being conducted simultaneously. The hydrants should be closed slowly to prevent problems resulting from surges. The number of hydrants to be opened depends on the pressure drop at

Fig. 8.24. Conducting fire flow and head loss test.

the residual hydrant. The change in pressure between the static condition and condition when hydrant is open must be at least 10 psi and should preferably be 20 psi.

With the above data, it is possible to estimate the discharge at some other residual pressure using the following formula:

$$Q_f = Q_t \left(\frac{P_s - P_f}{P_s - P_t} \right)^{0.54} \qquad (8.23.1)$$

where

Q_f = discharge at residual pressure P_f corresponding to fire condition, gpm
Q_t = discharge during test, gpm
P_s = residual pressure with no hydrants open, psi
P_t = residual pressure during test, psi
P_f = residual pressure during fire condition, psi.

Other units of pressure and flow can be used in the above equation as long as all pressures are expressed in the same units and all flows are expressed in the same units. The residual pressure during the fire is usually given as 20 psi for fire departments with pumper engines and 40 to 75 psi if water is used directly from hydrants, depending on the type of structures in the area.

The data collected in the above tests can be used to provide an indication of the distribution system's ability to provide water for fire fighting. The data can be more useful if two other pieces of information are also provided: (1) the head at the water source, and (2) the pressure drop through pressure reducing valves (PRV) for the static and test condition. These two additional pieces of information can be used along with the test results to provide: (1) better estimates of fire fighting capability and (2) data necessary for model calibration. The information on head at the source is easy to obtain, since records of tank water level and pump pressures are kept by most utilities. Pressure drop at pressure reducing valves can only be obtained by stationing an observer at each valve with two pressure gages or a differential gage and a way of communicating with those conducting the tests. An alternative to monitoring pressure at a PRV during the test is to simply bypass the valve during the test. This can be justified only if the setting on the PRV is such that it would indeed be open during a fire situation.

The ISO formula for estimating fire flow is derived below along with a more general equation. The derivations are presented in units of head in feet rather than pressure units since water elevation in tanks or head at pumping stations must be included. Both derivations are based on the assumptions that there is one source that provides most of the water to the test, and that the water use in the service area for non-fire purposes remains essentially constant during the test and hypothetical fire (i.e., $r = 1$ in Section 8.22). The head loss due to friction between the source and the residual hydrant can be given by the following equations for static, test and fire conditions:

$$h_s = Z_s - H_s - dP_s = KQ_s^n$$

$$h_t = Z_t - H_t - dP_t = K(Q_s + Q_t)^n \qquad (8.23.2)$$

$$h_f = Z_f - H_f - dP_f = K(Q_s + Q_f)^n$$

where
- h = friction head loss, ft
- Z = head at source, ft
- H = head at residual hydrant, ft
- dP = head drop across PRV, ft
- K = head loss coefficient for system represented by equivalent pipe
- Q_s = water use for non-fire purpose, gpm
- Q_t, Q_f = flow from hydrant for test and fire conditions respectively, gpm.

The subscripts on the head terms (s, t, f) refer to the static, test, and fire conditions, respectively. The quantities Z, H, dP, and Q_t are known or can be projected for the static, test, and fire conditions. There are three unknowns in the above equations: K, Q_s, and Q_f. K and Q_s can be eliminated from the above equations to give Q_f:

$$Q_f = Q_t \left[\frac{(h_f/h_s)^{1/n} - 1}{(h_t/h_s)^{1/n} - 1} \right]. \qquad (8.23.3)$$

To use Eq. (8.23.3), Z_s, Z_t, H_s, H_t, dP_s, dP_t, and Q_t are measured during the flow test, and n is taken as 1.85 (from the Hazen–Williams equation). Z_f, and dP_f are estimated based on the estimated head and pressure losses during a fire. If the source is a tank all the Z values will be equal, while if it is a pump the Z values will depend on the steepness of the head characteristic curve (although they may be virtually the same for fairly flat curves or when the fire flow is not excessive compared with the pumping capacity). The value of Z_f can also be changed to account for the fact that the tanks may be filled to a different level during a fire than during a test. The PRV should be completely open during the fire so dP_f should be zero if the valve is set correctly. H_f can be calculated based on the target pressure during a fire (e.g., 20 psi). With these data it is possible to calculate the h values to use in Eq. (8.23.3).

Equation (8.23.1) and (8.23.3) may not look very similar, but (8.23.1) is merely a simplified version of equation (8.23.3). The shorter equation incorporates two significant simplifying assumptions: (1) it ignores PRV effects (all $dP = 0$) and (2) it assumes that the head at the source is the same as the head at the residual hydrant during the static condition (all $Z = H_s$). These two assumptions are valid only in certain conditions. Eq. (8.23.1) should not be considered as a very accurate estimate of Q_f. Instead it can be viewed as a means of obtaining a rough indication of Q_f. Because of its simplicity, it will probably not be replaced in its role in evaluating water distribution systems for fire fighting purposes.

EXAMPLE. Water for fire fighting for an area comes from a tank with water level 300 ft. If the residual pressure at a hydrant at elevation 100 ft is 60 psi and 43 psi for the static and test conditions, respectively, what flow can be delivered for fire fighting to the area at 20 psi if 1000 gpm was delivered during the test? The PRV between the source and the test reduced the head by 15 ft during the static condition and 5 ft during the flow test. Calculate the fire flow using both Eq. (8.23.3) and (8.23.1) and $n = 1.85$.

First calculate all of the h terms (2.31 ft of water = 1 psi)

$$h_s = 300 - [100 + 2.31(60)] - 15 = 46.4 \text{ ft}$$

$$h_t = 300 - [100 + 2.31(43)] - 5 = 105.7 \text{ ft}$$

$$h_f = 300 - [100 + 2.31(20)] = 153.8 \text{ ft}.$$

Substituting into equation (8.23.3) gives

$$Q_f = 1000 \left[\frac{(153.8/46.4)^{0.54} - 1}{(105.7/46.4)^{0.54} - 1} \right] = 1625 \text{ gpm.}$$

Using Eq. (8.23.1) gives

$$Q_f = 1000 \left(\frac{60 - 20}{43 - 20} \right)^{0.54} = 1348 \text{ gpm.}$$

The difference in this example is 17%, with Eq. (8.23.3) being more accurate but still not perfect, since it is based on several simplifying assumptions.

The flow that can be delivered by the distribution system is only one of several factors involved in evaluating the water supply aspects of fire fighting. Other aspects include the distribution of hydrants, adequacy of storage, and adequacy of source. The flow which can be delivered to the fire is compared with the required fire flow to determine the adequacy of the overall system. Readers are referred to the publications of the Insurance Services Office (1980) for more details.

8.24. ASSESSING CAPABILITY FOR EXPANSION

Water use increases over time as new housing is developed or new industries move into town. The question that must be answered before a new user comes on line is whether the system will be able to provide sufficient water at adequate pressure. If the utility has a pipe network simulation model, the question can be answered fairly easily by simulating the increased flow in the vicinity of the increased use. Conducting a hydrant flow test in the area to check calibration of the model should be the first step in the procedure.

Even if there is no model of the network, the engineer can still obtain an appreciation of the impact of new users by conducting a fire flow test during a high-demand period and extrapolating the results using the methods given in the preceding section. There are a few minor differences in the approach, namely the engineer usually knows the desired use and is attempting to solve backwards for the pressure that will be provided at the increased use rate.

Since the pressure at the downstream side of a pressure reducing valve can be adjusted, the setting on the PRV can be considered a design parameter, limited only by the upstream pressure. Therefore it is better to use the downstream pressure setting at a PRV (if one controls pressure for the area) as the head at the source (H_s), since the PRV will be controlling pressure even during high-use periods.

Equation (8.23.2) can be solved for head in the vicinity of increased use with no pressure reducing valves to give

$$H_f = Z_f - (Z_s - H_s)\left\{\frac{Q_f}{Q_t}\left[\left(\frac{h_t}{h_s}\right)^{0.54} - 1\right] + 1\right\}^{1.85} \qquad (8.24.1)$$

where the variables are the same as in the previous section except that the subscript f refers to future additional use rather than fire flow. If pressure reducing valves are to be used, the Z terms should be replaced by $Z - dP$ terms, where the dP's are the pressure settings at the PRV's, which can be selected by the engineer.

If the system is not adequate to provide the additional flows at adequate pressure, the engineer must consider ways of meeting the need. These include: (1) elevated storage near the use area, (2) additional piping, (3) cleaning and lining mains, (4) additional pumping, or (5) conservation measures (e.g., increased charges for peak hour usage to motivate industrial user to provide in-plant storage).

EXAMPLE. Consider a system which has 56 psi pressure under static high-demand conditions and 45 psi when hydrants are open to flow 800 gpm. The head at the main pumping station is 150 ft above the elevation of the test hydrant. What will the pressure be at the residual hydrant if a new industrial user will require 1200 gpm? 2000 gpm? Use the elevation of the test hydrant as the datum.)

First calculate h_s and h_t:

$$h_s = 150 - 2.31(56) = 21 \text{ ft}$$

$$h_t = 150 - 2.31(45) = 46 \text{ ft.}$$

Substituting into Eq. (8.24.1) gives

$$H_f = 150 - (150 - 129) \left\{ \frac{1200}{800} \left[\left(\frac{46}{21} \right)^{0.54} - 1 \right] + 1 \right\}^{1.85}$$

$$= 88 \text{ ft} \quad \text{or} \quad 38 \text{ psi.}$$

At 2000 gpm the pressure would be reduced to 50 ft or 21 psi.

8.25. TESTING LARGE METERS

Water meters, especially those with moving parts, will eventually wear out, and because of this they should be tested periodically. Small meters can be "swapped out" and tested in a calibration lab. Because of their size and the difficulty of finding a replacement, large meters are generally tested in place.

Testing meters in place usually requires a pitot tube, although a properly calibrated clamp-on ultrasonic gage or insertable propeller gage may also be used. In all cases, it is necessary to find a sufficiently long straight section of pipe upstream or downstream of the meter so that the velocity profile is consistent with that required for such a gage. If the measurement must be made near a disturbance, it is best to make pitot traverses in more than one direction (e.g., top to bottom and left to right) to reduce the error due to atypical velocity profiles.

Buettner (1980) recommends that the meter test be conducted over a 5-minute test period, with the pitot tube read every 30 seconds. The velocity readings should be averaged and multiplied by the pipe factor (which should be determined separately), cross-sectional area (based on actual diameter corrected for the amount by which the pitot tube reduces the area) and duration of the test. A sample data sheet based on Buettner (1980) is shown in Table 8.5.

The formula for calculating the volume of water passing through the meter in a period of time t is

$$\text{Vol} = (A - dA) \text{ PF } v(0)_{av} t(60)$$

Table 8.5. Large Meter Test Results.

Location:
Date: Time:
Conducted by:
Meter size: Manufacturer:
Units on Meter:
Initial reading:
Final reading:
Difference:
Multiplier factors:
Volume used:

Pipe size: in. nominal
                 in. actual
Area: ft^2
Correction for pitot: ft^2
Units on differential pressure gage:
Specific gravity of manometer fluid (if applicable):
Pipe factor:

Reading	Time	Differential Gage Reading	Velocity (ft/sec)
1			
2			
3			
4			
5			
6			
7			
8			
9			
10			
11			
		Total:	
		Average:	

$$Vol = (A - dA) \; PF \; v(0)_{av} t (60)$$

$$=$$

$$=$$

Comparison with meter: % difference

Table 8.6.

Location: Industry 2
Date: 23 June 83   Time: 10:15 a.m.
Conducted by: Walski
Meter size: 16″   Manufacturer: Brand X
Units on Meter: gallons
Initial reading: 415,144
Final reading: 421,012
Difference: 5868
Multiplier factors: 1
Volume used: 5868 gal

Pipe size: 16 in. nominal
           16 in. actual
Area: 1.40 ft²
Correction for pitot: 0.02 ft²
Units on differential pressure gage: in.
Specific gravity of manometer fluid: 0.0061 air at 75 psig
Pipe factor: 0.85

where

$$\text{Vol} = \text{volume of water through meter, ft}^3$$
$$A = \text{actual area of pipe, ft}^2$$
$$dA = \text{correction to area for pitot tube, ft}^2$$
$$PF = \text{pipe factor}$$
$$v(0)_{av} = \text{time averaged centerline velocity, ft/sec}$$
$$t = \text{duration of test, min.}$$

Since the pitot tube is only accurate to 5%, the meter should not be considered inaccurate unless there is more than a 5% difference between the meter reading and the pitot tube results. If an excessive difference is detected, the meter should be replaced or repaired, although, as an interim measure the meter readings can be adjusted using a correction factor based on the results of the pitot tube test.

EXAMPLE. Consider a test conducted on a 16-in. turbine meter using an air-filled manometer at 75 psi line pressure. The pitot gage coefficient is 0.83 and the pipe factor is 0.85. During the test, the meter reading changed from 415,144 to 421,012 in 5 minutes. The readings from the manometer versus time are given and the other quantities are calculated as shown in Table 8.6.

Reading	Time (min)	Differential Gage Reading	Velocity (ft/sec)
1	0:00	2.8	3.2
2	0:30	3.5	3.6
3	1:00	3.3	3.5
4	1:30	2.6	3.1
5	2:00	2.4	3.0
6	2:30	2.1	2.8
7	3:00	2.6	3.1
8	3:30	3.1	3.4
9	4:00	3.0	3.4
10	4:30	2.9	3.3
11	5:00	2.8	3.2
	Total		35.6
	Average		3.2

$$\text{Vol} = (A - dA)\ \text{PF}\ v(0)_{av}\ t(60)$$

$$= (1.40 - 0.02)(0.84)(3.2)5(60)$$

$$= 1126\ \text{ft}^3 = 8423\ \text{gal}$$

Comparison with meter -30% difference

REVIEW QUESTIONS

1. What is wrong with using "literature values" for C factors when deciding whether to clean and line pipes?
2. Why are manometers not very useful for measuring gage pressure in water distribution systems?
3. Give two meanings for the word *transducer*.
4. What is the difference in accuracy between a Grade 2A and a Grade B pressure gage?
5. What is a good range for a pressure gage for use in a water distribution system in psi? in KPa?
6. What kind of errors can occur when static pressure is read from a tap on a customer's service line as opposed to a hydrant?
7. Why should a hydrant valve be completely open when a pressure reading is taken?
8. Give three uses for differential pressure measuring devices used in water distribution systems.
9. In determining the type of manometer fluid to be used for a given application, an engineer determines the specific gravity should be 8. Is there a fluid with such a specific gravity?
10. Name three pressures that are important in selecting a differential pressure measuring device. Why they are important?
11. How do pressure snubbers work?

12. What problems can occur when a snubber provides too much snubbing?

13. Venturi meters, orifice plates, and nozzles are referred to as "head measuring" devices. Why is this, and why is the range of such devices only on the order of 4 : 1?

14. If a venturi meter in a given pipe is sized properly for a given flow rate, should a venturi for the same pipe at a lower flow rate have a narrower, wider, or same size throat?

15. When can an elbow meter measure flows equally well in either direction? Can a venturi measure flows in either direction? Why?

16. When would a compound meter be selected over a single turbine meter?

17. Describe the advantages and disadvantages of using a clamp-on ultrasonic flowmeter for measuring velocity in an old pipe.

18. Why are doppler flowmeters not very useful for drinking water applications?

19. Discuss the differences in using a clamp-on flowmeter versus a device that must be inserted through a corporation cock. Compare the cost of tapping a pipe versus that of excavating, shoring, and repaving to reach the pipe.

20. Why is a differential gage used with a pitot tube rather than two pressure gages?

21. In plotting the velocity profile from a traverse to determine average velocity, why do the outer annular rings appear narrower than the inner ones?

22. What is the "pipe factor" and why is it useful in making flow measurements during a water audit?

23. Why is the correction to cross section for the area taken up by the pitot tube small for large pipes?

24. What are the advantages of a symmetric pitot tube?

25. How does a pitot gage used to measure discharge from a hydrant differ from a pitot tube used for flow in pipes?

26. Describe how x and y can be measured for the trajectory method of calculating discharge in a real street with curbstones and sloping roads.

27. Describe the error that can occur by assuming that the distance over which head loss occurs in a pipe is the same as the distance between the two hydrants along which head loss is measured.

28. If a "literature value" is given for a C factor, should that value be used with the nominal or actual diameter, assuming they are different?

29. Why are data read from contour maps with 20-ft contour intervals not very useful for hydraulic gradient tests?

30. Explain how the parallel pipe method for measuring head loss works. Why is it generally more accurate than a two-gage method? What can go wrong with the parallel pipe method?

31. If a C factor is calculated based on the actual length of pipe, will it be larger or smaller than a C based on the equivalent length of the pipe with minor losses represented by extra pipe?

32. The kinematic viscosity of water can vary by a factor of two between 40 and 90 deg F, yet this effect is not accounted for in the Hazen–Williams C. Why can it be ignored with little error?

33. Compare the methods for indirectly determining C using Eqs. (8.21.1) and (8.21.5). When is each more appropriate?

34. Why should h_2 be much greater than h_1 for Eq. (8.21.5) to be accurate?

35. What could cause a very steep slope in the hydraulic grade line?
36. Why is it better to conduct hydraulic gradient tests during high-water-use periods?
37. In using a model to assess the ability of a water distribution system to meet fire flows, why is it necessary to include greater detail (i.e., less skeletonizing) in the area around the fire demand?
38. Compare the advantages and disadvantages of Eq. (8.23.1) versus Eq. (8.23.3).
39. Why are pressure reducing valves generally set to be wide open during a fire flow situation? Compare using a PRV to using a smaller diameter pipe to reduce pressure.
40. Describe how a fire flow test can be used to assess the ability of a system to handle new users.
41. Why is it advantageous to test large meters in place?
42. When is it necessary to run a pitot traverse in checking large meter accuracy?

PROBLEMS

1. Determine the C required for an orifice plate in a 12-in. pipeline which carries between 2.4 and 0.8 mgd such that the differential pressure reading at the lowest flow is 1 in. of water. Given that C, what will the differential reading be at 2.4 mgd? (*Ans.* 0.68, 9.0 in.)

2. Determine the specific gravity required for a manometer fluid which will give a reading of 1 in. at a velocity of 1.24 ft/sec when used with a pitot tube with a coefficient of 0.83. Find the specific gravity of both a heavy and light fluid which can be used.

   Suppose the manometer is inclined 45 deg. What is the required specific gravity? (*Ans.* 1.415, 0.585, 1.587, 0.413)

3. Given a pitot tube with coefficient 0.83 used in a 10-in. pipe with an air-filled manometer and line pressure 90 psi, find the average velocity and pipe factor. The readings and the distance from the bottom of the pipe are given below. (*Ans.* 3.4 ft/sec, PF = 0.83.)

Distance from Bottom (in.)	Reading (in.)
0.5	2.3
1.	2.8
2.	3.4
3.	3.8
4.	4.1
5.	4.3
6.	4.0
7.	3.7
8.	3.3
9.	2.7
9.5	2.2

4. Calculate the discharge from a $2\frac{1}{2}$-in. hydrant outlet by the trajectory method assuming the vena contracta was $2\frac{1}{2}$ in. The stream travelled 22 ft as it dropped 3 ft. (*Ans.* 780 gpm.)

5. Determine the range on a pressure gage attached to a pitot gage so that it will still be on scale when the discharge is 1200 gpm from a $2\frac{1}{2}$-in. hydrant outlet. Use $C=$ 0.90. (*Ans.* 52 psi.)

6. For a head loss test conducted using the parallel pipe method to determine head and a pitot tube to measure velocity, determine the C factor and pipe roughness, given that the inside diameter is 15.1 in., the head loss in 600 ft of pipe was 3.2 psi, the velocity was 4.2 ft/sec, and the kinematic viscosity of water at the test temperature is 10^{-5} ft²/sec. What is the C factor if the nominal diameter of 16 in. is used? What would the pipe roughness be if the term containing the Reynolds number was ignored in the Colebrook–White equation? Why? (*Ans.* 71, 0.029, 68, 0.029, rough pipe.)

7. What is the C factor and water use in an area served by a 20-in. pipe for which the head loss is 0.9 ft with the test hydrant closed and 2.1 ft with the hydrant discharging 1200 gpm? Assume $r = 1$. Is the result based on nominal or actual diameter? (*Ans.* 112, 2064 gpm, nominal.)

8. Given the results of a hydraulic gradient test shown in the table below, plot the hydraulic grade line. Which section of pipe is responsible for the most head loss and what is the direction of flow in the last section? (*Ans.* Section 3, toward $x = 0$.)

x (ft)	z (ft)	p (psi)
0	0	208
7,040	54	61
12,200	91	42
16,560	38	40
21,130	74	30

9. Given the static and test pressure from a hydrant flow test of 65 and 56 psi, respectively, and a test flow of 800 gpm, find the discharge at 20 psi if the head at the source is 210 ft above the elevation of the hydrant for all cases and there are no pressure reducing valves. Use the ISO equation and the method given in Eq. (8.23.3). (*Ans.* 1908 gpm, 3274 gpm.)

10. During a 5-minute meter test the meter on a 24-in. line (actual diameter 23.3 in.) changed from 178,122 to 179,753. Given that the pipe factor was 0.91 and the pitot tube coefficient was 0.83, what is the difference between the pitot tube results and the meter reading for the 11 pitot tube readings taken at the centerline of the pipe at 30 sec intervals shown below? The specific gravity of the manometer fluid was 1.2. (*Ans.* 1.4%.)

Manometer
Reading
(in.)

4.2
4.1
4.3
4.7
4.8
4.8
4.8
4.3
4.3
4.7
4.3

REFERENCES

American National Standards Institute, 1974, "Gages, Pressure and Vacuum Indicating Dial Type-Elastic Element," ANSI B40.1-1974.

Buettner, C.F., 1980, *Practical Hydraulics and Water Flow Monitoring Workshop*, Manual, Box 8656, St. Louis, MO.

Cole, E.S., 1935, "Pitot Tube Practice," *Transactions ASME*, Vol. HYD-57, No. 8, p. 281.

Cole, E.S., and E.S. Cole, 1939, "Pitot Tubes in Large Pipes," *Transactions ASME*, Vol. 61, p. 465.

Fath, J.P., 1978, "Water Flow Measurement in Large Pipes and Conduits," *CTI Annual Technical Meeting*, Houstin, TX, TP-182A.

Guthrie, D.L., 1981, *NPDES Compliance Flow Measurement Manual*, MCD-77, U.S. EPA, Washington, D.C.

Hauserman, W.R., 1980, "A 250-year Old Flow Meter—The Pitot Tube," *Instrument Society of America, International Conference*, Houston, TX, Vol. 1, p. 339.

Hudson, W.D., 1954, "Flow Tests on Distribution Systems," *J. AWWA*, Vol. 46, No. 2, p. 144.

Insurance Services Office, 1973, *Fire Flow Tests*, New York, NY.

Insurance Services Office, 1980, *Fire Suppression Rating Schedule*, New York, NY.

Miller, R.W., 1983, *Flow Measurement Engineering Handbook*, McGraw-Hill, New York.

Replogle, J.A., L.E. Myers, and K.J. Brust, 1966, "Evaluation of Pipe Elbows as Flow Meters," *J. ASCE Irrigation and Drainage Div.*, Vol. 92, No. IR 3, p. 17.

Robertson, J.M., and M.E. Clark, 1977, "On Improving the Pitot-Tube Determination of Flows in Large Pipes," National Bureau of Standards Special Publication 484, Gaithersburg, MD.

APPENDIX.
CONVERSION
FACTORS

Flow (volume/time).

	m³/sec	mgd	cfs	gpm	ac-ft/day
m³/sec	1	22.8	35.3	15850	70.0
mgd	0.0438	1	1.55	694	3.07
cfs	0.0283	0.646	1	448	1.98
gpm	6.31×10^{-5}	0.00144	0.00223	1	0.00442
ac-ft/day	0.0143	0.326	0.504	226.	1

mgd = million gallons per day
cfs = cubic feet per second
gpm = gallons per minute
ac-ft/day = acre-ft/day
m = meter
(Example: To convert 5 mgd to gpm, 5 mgd × 694 gpm/mgd = 3470 gpm)

Kinematic Viscosity

	m²/sec	cm²/sec	ft²/sec	centistoke
m²/sec	1	10^4	10.7	10^6
cm²/sec (stoke)	10^{-4}	1	0.00107	100
ft²/sec	0.0929	929	1	9.34×10^4
centistoke	10^{-6}	0.01	1.07×10^{-5}	1

cm = centimeter

Pressure (force/area).

	kPa	psi	ft H_2O	m H_2O	atm
kPa	1	0.145	0.334	0.102	0.00989
psi	6.89	1	2.31	0.704	0.0680
ft H_2O	2.99	0.433	1	0.305	0.0294
m H_2O	9.81	1.42	3.28	1	0.0965
atm	101	14.7	33.9	10.4	1

1 kPa = 1 kN/m²
kPa = kilopascal
kN/m² = kilonewton per square meter
psi = pounds force per square inch
ft H_2O = feet of water column which would exert the same pressure
m H_2O = meters of water column which would exert the same pressure
atm = atmospheres

To convert from feet or meters of water to similar units of another fluid divide by the specific gravity of that fluid (e.g., to convert ft water to ft mercury divide by 13.6, the specific gravity of mercury.)

Power (energy/time).

	joule/sec	ft-lb$_f$/sec	kilowatt	hp	Btu/sec
Joule/sec (watt)	1	0.738	0.001	0.00134	9.48×10^{-4}
ft-lb$_f$/sec	1.36	1	0.00136	0.00182	0.00128
kilowatt	1000	738	1	1.34	0.948
horsepower (hp)	746	550	0.746	1	0.707
Btu/sec	1055	778	1.05	1.41	1

lb$_f$ = pound force
Btu = British thermal unit

Length

	m	ft	in.	mm	mi	km
m	1	3.28	39.4	1000	6.21×10^{-4}	0.001
ft	0.305	1	12.0	305	1.89×10^{-4}	3.05×10^{-4}
in.	0.0254	0.0833	1	25.4	1.58×10^{-5}	2.54×10^{-5}
mm	0.001	0.00328	0.0394	1	6.21×10^{-7}	10
mi	1609	5280	6.34×10^4	1.61×10^6	1	1.61
km	1000	3280	3.94×10^4	10	0.621	1

in. = inches
mm = millimeter
mi = mile
km = kilometer

INDEX

Blasius equation, 29
boundary layer, 21, 30–32
Bourdon gage, 212
branched network, 51

C-factor. *See* Hazen-Williams
Caldwell-Lawrence diagram, 151
cathodic protection, 179, 191–194, 199–200
cleaning and lining, 149–176
  comparison with cleaning only, 175–176
  comparison with energy cost, 160–166
  comparison with parallel pipe, 166–175
  costs, 159–160
  economic evaluation, 155–176
  process, 153–156
Colebrook-White equation, 30–33, 250
consumer surplus, 198
continuity equation
  differential form, 9–11
  integral form, 6–9
corrosion, 149, 180–181
critical break rate, 203–204

Dahl tube, 226
Darcy-Weisbach equation, 25–35, 51, 96
differential pressure gage, 214–219
  for head loss test, 247–248
  for pitot tube, 232–234
discharge measurement, 240–242
distributed pitot tube, 239
Doppler meter, 231

eddy viscosity, 17
efficiency, 64–66
elbow meter, 226–228
electromagnetic flowmeter, 231

elevation head, 19
energy equation, 17–20
energy grade line, 19
energy gradient, 19
energy head, 19
equivalent pipes, 52–56
  parallel, 54–56
  series, 52–54
extended period stimulation, 66–69

fire flow tests, 257–261
flow equations, 73–74, 77–80
flow measurement, 220–242
flow tubes, 226
flow units, 3, appendix
friction factor, 22–35
  laminar flow, 26, 28
  turbulent flow, 27–35
friction velocity, 21–22
graphitization, 181

Hagen-Poiseuille equation, 16, 26
Hardy-Cross method, 49, 76–77, 83–85
Hazen-Williams equation, 34–42, 51, 96
  C-factor decrease, 150–153
  C-factor table, 35–36
  C-factor tests, 247–255
head characteristic curve, 55–64
head loss measurement, 247–249
head loss nomogram
  rough flow, 43
  smooth flow, 38
horsepower
  brake, 64
  water, 64
  wire, 65

hydrant flow test
  description, 240–242, 257–263
  for model calibration, 99–103
hydraulic grade line, 19
hydraulic gradient, 19
hydraulic gradient tests, 255–257
hydrostatic testing, 186–187

kinematic viscosity
  conversion factors, appendix
  of water, 250
kinetic energy correction factor, 19–20

Law of the Wall, 23
leak detection, 179, 182–194
  evaluation, 187–190
linear theory method, 49, 76–80
listening, 182, 184–186
loop equations, 75–76, 83–85
looped network, 50–52, 70–85
lost water, 179, 206–207

Manning equation, 34
manometer, 212, 215–218
maps, 96–97
minimum night ratio, 183–184
minor losses, 42–45, 248–249
models. *See* water distribution models
momentum equation
  correction factor, 12
  differential form, 13–17
  integral form, 11–13
Moody diagram, 27–29, 250

Navier-Stokes equation, 13–14
network
  branched, 51
  definition, 49
  looped, 50–52, 70–85
  problems, 49–85
Newton-Raphson method, 49
  friction factor, 32
  network problems, 76–77, 80–83
  pump problems, 61–62
Newton's Law of Viscosity, 20–21
node equations, 74–75, 80–83
nozzle, 221–226

One-seventh Power Law, 23–34
orifice plate, 221–226

parallel pipe method, 247–248
piezometer head, 19
pipe breaks, 179–206
  causes, 180–182
  cost, 196–200
  economic evaluation, 200–204
  prediction, 194–196
  record keeping, 204–206
  units, 194
pipe caliper, 243–245
pipe costs, 119–122
pipe diameter, 243–246
pipe factor, 237
pipe repair, 191–194
pipe size selection, 117–147
  gravity, 127–136
  mathematical programming, 142–143
  nomogram, 136–142
  pumped, 122–127
pitot gage, 240–242, 258
pitot rod. See pitot tube
pitot tube, 183, 231–239
Power Law for Velocity, 23, 235–237
pressure gages, 212–214, 258
pressure head, 19
pressure snubbers, 219–220
pressure transducer, 212
pressure units, 3, appendix
pseudo-loop, 50, 70, 72
pumps, 56–66
  efficiency, 64
  parallel, 62–64

Reynolds equation, 16–17
Reynolds number, 15, 26
rough flow, 22, 27, 29
  nomogram, 43
roughness of pipes, 31, 247–255

scaling, 149
Scobey equation, 34
sliplining, 156
smooth flow, 22, 27, 29–30
  nomogram, 38
static head, 19
Swamee and Jain formulas, 33, 37
system head curve, 56–64
trajectory method, 240
turbulent core, 21

ultrasonic meters, 229–231
unaccounted-for water, 170, 206–207
unit conversions, 3, appendix

Velocity Defect Law, 23
velocity head, 19
venturi meter, 220–221
von-Karmen-Nikuradse equation, 29
vortex shedding meter, 229

water audit, 182, 188–189
water conditioning. *See* water stabilization

water distribution models, 91–115
   calibration, 98–113, 259
   computer program selection, 93–96
   extended period simulation, 93–94
   optimization, 93–94, 142–143
   production runs, 113–115
   simulation, 93–94
   skeletal system, 97
   steady state, 93–94
water meter testing, 263–266
water stabilization, 148, 151–153
water system expansion, 262–263
water use
   data for modeling, 98–102
Wood formula, 32–33